The Elements of Geography in Colour

SECOND EDITION

F.R. Dobson and H.E. Virgo

Second edition prepared by David P. Jones

HODDER AND STOUGHTON
LONDON SYDNEY AUCKLAND TORONTO

The Elements of
Geography in Colour

F. R. Dobson, BA
Formerly Senior Geography Master
The Forest Grammar School
Winnersh, Berkshire

AND

H. E. Virgo, MA
Formerly Vice Principal
The John Cleveland College
Hinckley, Leicestershire

New material for the second edition
prepared by David P. Jones

The map extracts on pages 16, 17, 20, 21, 24, 25 and 28 are reproduced with the sanction of the Controller of H. M. Stationery Office, Crown Copyright reserved.

Cover photographs and the photograph on page 29 were taken by D. Jones

ISBN 0 340 38590 1

First published 1974
Second edition 1987

Typeset by Macmillan India Ltd., Bangalore 25

Printed in Hong Kong
for Hodder and Stoughton Educational,
a division of Hodder and Stoughton Ltd,
Mill Road, Dunton Green, Sevenoaks, Kent
by Colorcraft Ltd

Debbie Barrett.
22-10-87

Preface to the Second Edition

This new edition of *The Elements of Geography in Colour* retains the long established and distinctive framework and format of the first edition. Throughout the book factual content and statistics have been updated to take account of the many changes in the world. In addition, new topics have been introduced reflecting some of the changes of interest within geography itself, so that hydrology, for example, is now covered. The Ordnance Survey map extracts have been replaced with the most recent editions and the accompanying text and questions have been amended as necessary. Exercises throughout the text have also been revised to reflect the criteria of the GCSE. Part Five has been rewritten to include a chapter on population and development as well as dealing with industry and resources.

The book provides a sound and interesting course for pupils in the fourth and fifth year preparing for external examinations, as the content is appropriate to the majority of new syllabuses. Like the previous edition Part One introduces map reading, a central skill in geography. Parts Two and Three still deal with the physical basis of geography and Part Four examines the principal natural regions of the world. In the context of different natural regions a wide variety of human activities, issues and problems are studied. A major change in Part Five extends the book's coverage to include population and development, which, with industry and resources, completes the book.

There are questions and exercises throughout the book which are intended to reinforce the work and to make pupils observe, think and investigate. The material in the book provides much scope for teachers who wish to devise their own exercises.

David P. Jones

Contents

Part One *Map Reading*
1 An Introduction to Map Reading 2
2 A Geographical Background to Map Reading 8
Exercises in Map Reading 11

Part Two *The Earth and its Atmosphere*
3 The Earth and its Movements 30
4 The Atmosphere 38
5 The Measurement of Climate and Weather 47

Part Three *The Surface of the Earth*
6 The Earth's Crust and Earth Movements 58
7 Earth Sculpture in Humid Regions 70
8 Earth Sculpture in Arid Regions 82
9 Glaciers and Ice Sheets 88
10 The Work of the Sea 96
11 Oceans and Seas 103

Part Four *The Natural Regions of the World*
12 Equatorial Regions 110
13 Savanna Regions 118
14 Hot Deserts 126
15 Monsoon Lands 131
16 Warm Temperate Western Margins (Mediterranean Regions) 140
17 Warm Temperate Eastern Margins 146
18 Temperate Interiors 151
19 Cool Temperate Western Margins 157
20 Cool Temperate Eastern Margins 164
21 Cold Temperate Regions 170
22 High Mountain and Tundra Areas 175

Part Five *Population and Industry*
23 Population and Development 182
24 Industry and Resources 188

Index 200

PART ONE

Map Reading

Chapter 1

An Introduction to Map Reading

In order to gain most benefit from the first two chapters of this book pupils are strongly advised to purchase their local 1 : 50 000 Ordnance Survey map. Ordnance Survey maps are published to cover all parts of the country. Prominent surface features, both natural and man-made, are marked on them by easily recognisable symbols known as conventional signs. The signs used on the 1 : 50 000 and 1 : 25 000 maps are given on pages 12 and 13.

SCALE

The scale of a map shows the relationship between any length on the map and the horizontal distance on the ground. It can be expressed in three ways:
(1) In words; for example '2 centimetres to 1 kilometre'.
(2) By a line, divided into a number of equal parts, on which are marked the distances. One of these parts may be subdivided to show smaller distances.
(3) As a ratio; for example 1 : 50 000, which means that one centimetre on the map represents 50 000 centimetres (or $\frac{1}{2}$ kilometre) on the Earth's surface.

The Ordnance Survey employs a variety of scale in its maps. Very detailed plans required by town planners, engineers and others are published on scales of 1 : 10 560, 1 : 2 500 and, in some cases, 1 : 1 250. However, easily the most popular are the 1 : 50 000 and 1 : 25 000 maps. These show many details and are suitable for anyone who wishes to explore an area on foot.

Measuring Distances on Maps

A ruler or pair of dividers is used to measure distances in a straight line on a map. For distances along winding roads or rivers a straight edge of paper is needed. After marking the starting point, twist the paper to follow the curves, using a sharp pencil or pin point to prevent it from slipping. Finally, mark the end and place the edge of paper along the scale line so that whole kilometres are measured to the right of the zero mark and the remaining portion of a kilometre to the left, in metres. Another method is to use a small instrument called an *opisometer* or map measurer which has a milled wheel that can be run along a road, the distance being indicated on a dial.

DIRECTION

Direction is expressed by means of the compass points, north and south being the directions of the North and South Poles. It may also be expressed in degrees as a three-figure bearing, being the angle measured from north in a clockwise direction.

A circular compass card 15 or 20 cm in diameter can be cut out of cardboard or hardboard. The compass points and degrees, as shown in figure 1.1, should be marked round the edge of the card and an empty cotton reel with a pencil through it should be fixed exactly over the centre. The card may now be placed so that the Sun's rays fall on it. At noon (GMT) turn the card until the shadow of the pencil falls on the north line. The card is now set and may be used to find the direction (bearing) of any object by looking from behind with both pencil and object in line and noting where the line of sight cuts the far edge of the card.

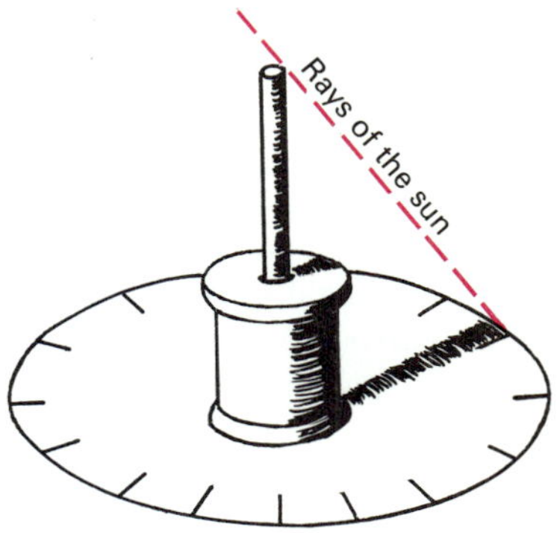

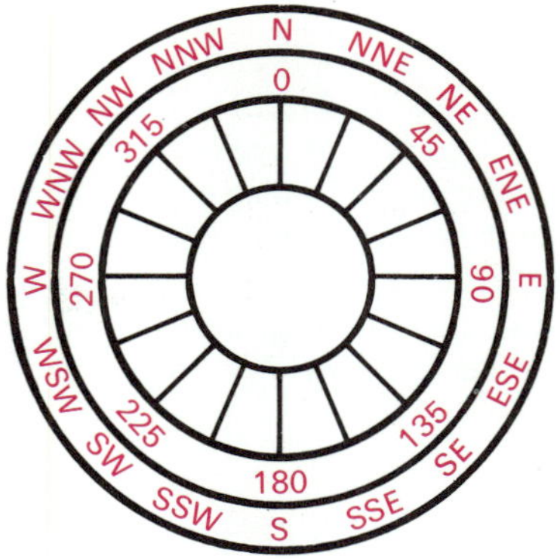

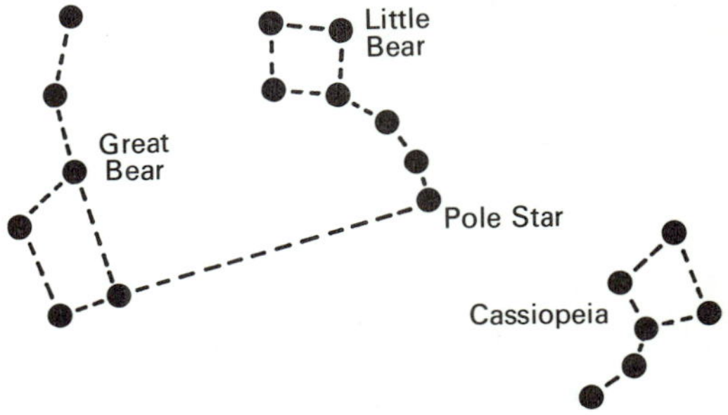

Figure 1.1 Compass card and points of the compass

Figure 1.2 How to find the Pole Star

True North

True north is the direction of the North Pole from any place on the Earth's surface. On atlas maps the lines of longitude all point true north and south. On Ordnance Survey maps the north–south grid lines run approximately true north to south. Since all the grid lines are parallel, there is a slight difference between grid north and true north, the amount being stated in the margin of each map.

Magnetic North

The Earth behaves as a huge magnet, but its North and South Magnetic Poles vary slightly in position from year to year, and do not coincide with the geographic Poles. The Magnetic North Pole lies north-north-west of the Boothia Peninsula in North America, and the Magnetic South Pole is in Victoria Land, Antarctica. A magnetic compass needle, if allowed to swing freely in a horizontal plane, comes to rest pointing towards magnetic north. The angle between true north and magnetic north is called the *magnetic variation*. Ordnance Survey maps state the amount of this variation, with the date and annual rate of change. In 1985 magnetic north over the British Isles was $5\frac{1}{2}^\circ$ west of true north.

Methods of Finding North and South

The most convenient way of finding north is by means of a magnetic compass. There are many types, ranging from the elaborate prismatic compass fitted with sights to give three-figure bearings on an object, to a number of simple, reliable and inexpensive ones. A magnetic compass is essential for any serious outdoor geographical work, but the user must remember to make allowance for the magnetic variation when true bearings are required.

If one is without a magnetic compass it is possible to find north and south by means of the Sun and stars. By night one can make use of the *Pole Star* which is approximately north of any observer in the northern hemisphere. Its position is shown in figure 1.2 in relation to the *Great Bear* and *Cassiopeia*, two easily recognisable constellations. By day the Sun can be used. In Britain the shadow of a vertical pole points north at noon (GMT). Alternatively a watch may be used. Point the hour hand to the Sun and bisect the angle between the hour hand and twelve o'clock: the bisector points due south. The watch method is not very accurate.

Setting a Map

Spread out your local Ordnance Survey map out of doors or near a window commanding a good view. Place a magnetic compass against the magnetic north sign and turn the map slowly until the

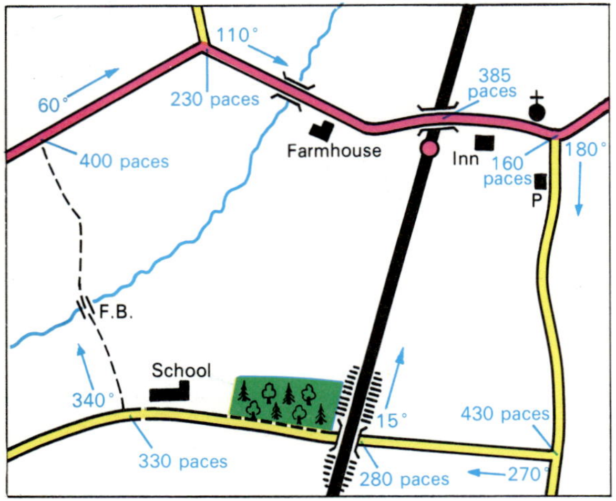

Figure 1.3 A compass traverse

compass needle points the same way as the magnetic north sign. The map is now set and it should be possible, provided that you know your exact position on the map, to identify various features such as roads, churches, etc., by taking bearings on them. This procedure is sometimes used by experienced hill walkers or climbers to enable them to find their way across difficult or trackless country.

A Compass Traverse

With the aid of a magnetic compass a simple map of a walk can be made. Choose a route two or three km in length. It must be divided into fairly straight stretches or 'legs'. At the starting point use your compass to find the direction of the first leg and mark this on your paper. Then pace the distance and find the direction of the next leg. Make a rough sketch map as you go along, entering the number of paces and the new direction at the end of each leg. Where you cross a railway or stream make a note of the distance and its direction. Fill in any details such as the type of road, houses, orchards or other features adjoining or leading off the route. Finally decide on the most suitable scale (bearing in mind that your pace is probably rather less than one metre) and draw the map neatly, using 1 : 50 000 Ordnance Survey signs. Add the compass points and scale.

RELIEF

Relief is the variation in height of the Earth's surface. Mountains, valleys, plateaus and plains are features of relief. On early maps relief was frequently shown by *hachures*, which are a series of lines running down the slopes, the lines being closer and thicker where they are gentle. Another method is known as *hill shading* in which it is assumed that a bright light, shining obliquely (and generally from the north-west), illuminates the area. The differences of light and shade reveal the shape of high and low land. Both hachures and hill shading give an attractive pictorial representation of the features but do not show accurately the height of the land.

Contours are usually employed to show relief in modern maps. These are lines joining places of the same height above mean sea level. For Britain, mean sea level is taken as the average level of the sea at Newlyn, Cornwall. The difference in height between contours is termed the *vertical interval*. In addition to showing the height of the land, contours indicate the steepness of slope at any place, being closer together where the slope is steep and farther apart where it is gentle. To show the exact height in certain places such as mountain summits, *spot heights* are used to supplement the contours. These are marked by a dot, with the height in feet. A triangle round the dot means that it is also a *triangulation station*. These are used from time to time by the Ordnance Survey when re-surveying the area.

In order to make contour maps easier to read a system of *layer tinting* is often employed, especially on atlas and tourist maps. Areas between the contours are coloured, the lowlands being either uncoloured or in green, higher land in yellow, followed by light brown, dark brown, purple and perhaps white for mountain summits.

The Representation of Land Forms by Means of Contours

Three areas are shown in figures 1.4 and 1.5, containing between them some of the commonest types of land form. Each area is shown both as a simple line drawing and as a contour map. The features marked on the drawings should be identified on the maps by means of the shapes and patterns made by the contours. Some definitions of the land forms shown in the three areas follow:

Hill or Mountain. An area of high land, often rising to a peak. In Britain an upland rising over 500 metres above the surrounding land is usually referred to as a mountain.

Summit. The highest point of a hill or mountain. If the summit has steep sides it is known as a *peak*.

Mountain Range. A line of mountains.

Plateau. A flat area of high land.

Ridge. A long, narrow area of high land.
Spur. A long, narrow projection of high land into lower land.
Knoll. A low, detached hill.
Cuesta. A ridge with a steep slope on one side (known as the scarp slope or escarpment) and a gentle slope on the other (dip slope).
Valley. A long, narrow depression, usually occupied by a river.
River Basin. The total area drained by a river and its tributaries.
Watershed. A divide between two drainage areas, separating streams flowing into one river basin from those flowing into another.
Col or Saddle. A dip between two summits.
Gap or Pass. An easy way through high land.
Plain. A flat area of low land.
Promontory or Headland. High land jutting out to sea.
Cliff. A precipitous slope.
Estuary. A wide river mouth.
Bay and Cove. A large and a small coastal inlet.
Convex Slope. A slope that is steep at the bottom, becoming less steep higher up.
Concave Slope. A slope that is steep at the top, becoming less steep lower down.

Cross-sections and Visibility

To draw a cross-section, first draw a straight line across a contour map (see figure 1.6). Construct a frame consisting of a base line at least as long as the line on the map, with a vertical line between 2 and 5 cm high at the left side. Mark heights, at suitably spaced intervals, on this vertical line to include all the heights shown along the line on the map. Draw

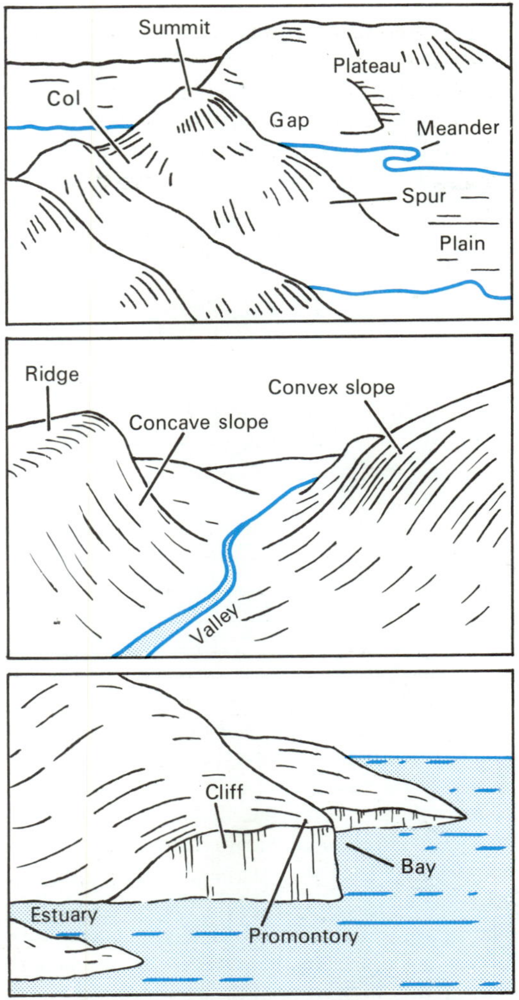

Figure 1.4 Some typical land forms

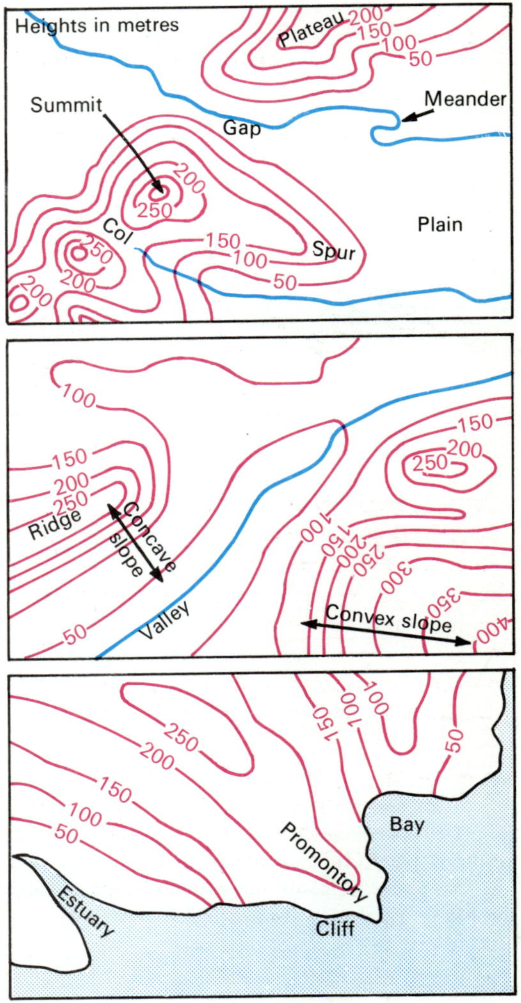

Figure 1.5 Their contour equivalents

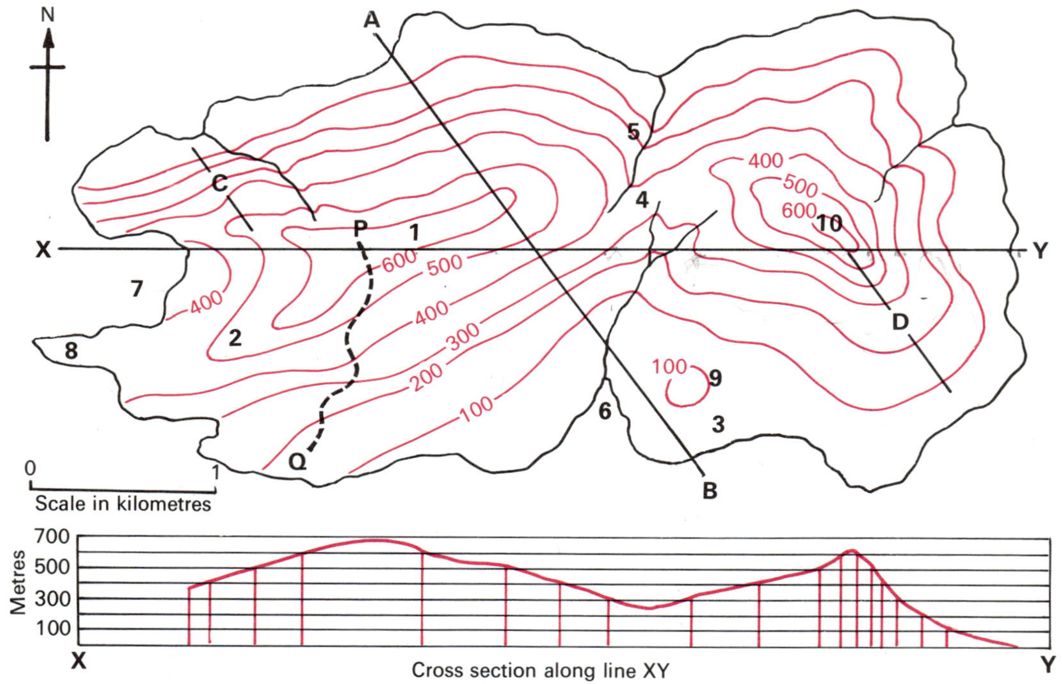

Figure 1.6 A contour map of an island
(*a*) What is the greatest distance across the island from east to west and from north to south? (*b*) What vertical interval has been used for the contours? (*c*) Each number from 1 to 10 represents a common type of land form. What are they? (*d*) Draw a cross-section from A to B. (*e*) What types of slope are found at C and D? (*f*) What is the average gradient of the footpath from P to Q?

horizontal lines from these height marks. Place a straight edge of paper along the line on the map and mark with a dash where each contour crosses or touches the line. Against each dash make a note of the height. Mark in also any streams, coastlines or spot heights. Then place the edge of paper along the base line of your frame. Make dashes on the base line opposite to those on your paper and draw perpendiculars from these dashes to the heights indicated. Join the tops of the perpendiculars, with a curved line where necessary, to complete the cross-section.

Cross-sections show the shape of the land and help in deciding whether one point on the ground is visible from another. If the slope is uniform the bottom of a hill will just be visible from the top, provided that there are no obstructions such as trees or buildings. If the slope is convex the bottom will not be visible from the top; on the other hand, if it is concave the whole of the base of the hill will be visible. Many hill slopes are a mixture of these types, whilst the surfaces of plains or plateaus may be undulating (i.e. gently rising and falling). In such cases it may be difficult to decide problems of visibility except by the drawing of an accurate cross section.

Gradient

Gradient is the degree of slope. The average gradient between any two points, whether 'measured' in a straight line or along a road or river, may be calculated by dividing the distance apart of the two points by the difference in height (both in metres). Thus if the distance apart is 2000 metres and the difference in height is 200 metres, then the gradient is 1 in 10. This is more often expressed as a percentage now, especially on road signs. In this example it would be 10%.

Location

Where is Nairobi? Look it up in the index at the back of your atlas. Against the name you will find the page number together with its latitude and longitude in degrees and minutes (a minute is a sixtieth of a degree). Latitude is measured north or south of the Equator and longitude is measured

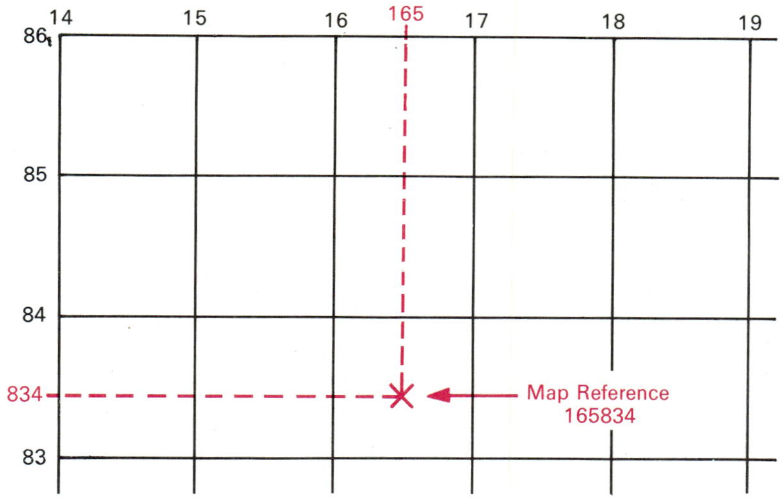

Figure 1.7 A six-figure map reference

east or west of the Greenwich Meridian (the line of longitude that passes through Greenwich). The degrees of latitude will be found at the sides of a map and the degrees of longitude at the top or bottom.

Ordnance Survey maps do not employ latitude and longitude, but have their own '*national grid*' made up of parallel lines enclosing squares whose sides measure one kilometre. The lines running from top to bottom are known as *eastings* and those running from side to side as *northings*. Six-figure *map references* are used to find places on the maps. An example will help to explain this. In the map reference 165834 the first three figures represent the easting and the last three the northing. An easting of 165 means that the point lies midway between the eastings numbered 16 and 17, that is on the vertical dotted line in figure 1.7. Similarly the northing 834 is represented by the dotted line four-tenths of the way between the northings numbered 83 and 84.

Exercises

1 Draw a square with sides 15 centimetres long. Inside the square draw a map of either your school or your home area, using Ordnance Survey symbols. Beneath the map indicate the scale in three different ways and show the north by means of an arrow.

2 **Making a Relief Model**
An excellent way to improve one's understanding of contours is to make a relief model of one of the contour maps given in this book. First enlarge the map to a suitable size, say 30 × 40 centimetres. Make separate tracings of each contour on cardboard, hardboard or thin plywood and cut them out with large scissors or a fretsaw. Glue or screw the pieces together in the correct order and position on top of a baseboard. The sharp edges represent the contours and the steps may be filled in with Plasticine, modelling clay or other suitable material. The contours should be carefully drawn in and, if possible, layer colouring applied and streams marked.

An alternative way of making the model is to use papier mâché on top of a base board. Small pieces of newspaper are boiled with glue and a small amount of plaster of paris. From this a very hard-wearing model can be made, the surface of which may be sandpapered for a neater finish. Flour paste can be substituted for glue, though it is less suitable.

Chapter 2

A Geographical Background to Map Reading

We can learn a great deal about the human activities in any area by a study of a local Ordnance Survey sheets. Houses and places of work (factories, farms, mines, quarries, business offices, etc.) are marked, together with the means of communication, including roads, railways, waterways and airports. By working out the relationship between these various aspects of human life and the physical geography (the relief, drainage, rock type and natural vegetation) it is possible to build up a picture of the geography of the area as a whole.

SETTLEMENT

Many modern settlements have their roots deep in the past. The earliest sites marked on Ordnance Survey maps belong to the prehistoric (i.e. pre-Roman) periods, which include the Neolithic, Bronze and Early Iron Ages. The various remains of these periods are shown on the maps in early English (Gothic) lettering. They include British or Celtic villages, Celtic fields, lynchets (cultivation terraces), tumuli (burial mounds), hill forts (which are often called camps or castles), stone circles, ridgeways and other ancient trackways. Prehistoric people seem to have largely avoided the lowlands, which were covered with thick forest and marsh, preferring the more open upland sites.

Roman remains, which date from the first to the fifth century A D, are indicated by Roman lettering, and their distribution shows that the clearance of forest and drainage of marsh had started in earnest. Towns grew up at bridging points of rivers, at river mouths and at the meeting points of Roman roads. Many of these early settlements are now modern cities, for example *Londinium* (London), *Deva* (Chester) and *Eboracum* (York). In country districts, particularly in the south and east of England, were large numbers of Roman villas, many of which may have served as farm-houses.

The next period, which lasted from the fourth to the eleventh century A D, saw the occupation of the English lowlands by the Angles and Saxons and of parts of northern England and Scotland by the Norsemen. These peoples travelled mainly by water and chose fertile riverside sites on which to settle. Their early villages were almost self-supporting and the land included, if possible, a stretch of hillside for grazing sheep, some valley meadow for cattle and an area of gently sloping land for crop growing. Most English villages of today date from this period. Some retain their Saxon place names, with endings such as *den*, *ton*, *ing*, *ford*, *bury*, *ley* and *field*. Modern parish boundaries frequently run from the rivers right up to the crest of the hills and thus include all three types of land, suggesting that the villages may have retained their original boundaries. This type of settlement is generally compact, with the older houses, church and inn close together near a stream or spring, and is called a *nucleated village*.

A later form of village developed in the forested lowlands, occupying newly made clearings where water was easily obtainable from shallow wells and streams. The cottages often straggled along winding cart tracks which followed the field boundaries. Today these villages have generally retained their straggling pattern and the cart tracks have become the village roads.

Walton, Cumberland (*Aerofilms*)
Describe the scene. What type of village is this?

Some villages grew into *market towns*, where farmers from the surrounding countryside sold their surplus products. Villages with a geographical advantage, for example those at a ford or bridging point of a river, or where there was a stream which could work a mill, grew most rapidly. Manufacturing, which included the spinning and weaving of wool, the tanning of leather and metal-working, developed in some of these towns. Others occupied ~ good defensive site and spread out beneath the walls of a Norman castle. (Norman and medieval buildings are marked on Ordnance Survey maps in Gothic script.)

Today few towns retain the character of a true market town. Many of them have grown into *industrial towns*, with the market largely replaced by retail shops, whilst some, including the *county towns*, are administrative centres. With the increasing population the problem of housing more and more people became acute, and large blocks of flats were constructed in or near the towns. The building of new houses spread, and is still spreading, into the countryside, and in many places ribbon development has taken place along existing roads, leaving open country between. Large *housing estates*, on the other hands, are usually off the main roads and have their own service roads. In the case of very large cities, such as London, residential areas called *suburbs* have grown up in the outskirts. Farther away still are separate *commuter towns* from which many people travel daily to work in the main centre.

Many towns along the coast came to depend on sea-borne trade. Those with a sheltered harbour, usually on an estuary and with sufficient depth of water, flourished, whilst some decayed as the size of ships increased. In the 1950s and 1960s many coastal holiday resorts grew into considerable towns, especially those with a good sandy beach, shelter from strong winds and a sunny climate.

The actual choice of a settlement site would clearly depend on the local geographical conditions, and the following factors would tend to favour the growth of a settlement:

(1) A reliable supply of pure water.
(2) Gently rising ground, above the flood level. Villages in very low-lying areas often occupy low mounds, consisting of a gravel.
(3) A firm foundation, preferably of solid rock. Clay is not very suitable, since it swells when wetted and shrinks when it dries.
(4) Local building materials, for example clay, timber, stone and straw.
(5) Some protection from strong winds. From this point of view the shelter provided by hills is ideal.
(6) Fertile soils, local mineral wealth or forests to provide employment.
(7) Good routeways, on firm ground but without steep gradients.

It is true that many of these factors are no longer important. Building materials can now be conveyed from distant places, water can be piped from elsewhere, and the advent of the railway and motor car make local employment no longer necessary. Owing to the high density of population in many parts of the United Kingdom many people are compelled to live in areas which are not ideally suited geographically.

LAND USE

The exact use of the land is not usually indicated on Ordnance Survey maps. It is possible, however, after a careful study of a map, to work out the likely forms it may take.

In rural areas a large number of farms and cottages to the square kilometre suggests that the soil is fertile and that *intensive methods* of farming are employed. This means that the land is farmed carefully and a considerable amount of labour may be employed, so as to obtain the maximum yield per hectare. It may take the form of mixed farming (the growing of crops and keeping of livestock on the same farm), market gardening (the growing of vegetables and flowers, sometimes in hothouses), fruit growing or, on moist lowlands, dairying on permanent pasture. These types of farming are mainly restricted to plains and the lower slopes of hills. Very flat land may be artificially drained, with drainage ditches (distinguishable from streams by their straightness and right-angled bends) and dykes to prevent the rivers flooding.

In contrast there may be very few farms or cottages, and the fields may be larger, suggesting that the land is less fertile. Larger farms are needed, with less labour, if a farmer is to make a living, and *extensive methods* are generally employed, which means that less attention is given to each hectare. Examples of this are the highly mechanised arable farms in chalk and limestone areas, producing cereals, especially barley, and in mountain areas the huge sheep farms which cover thousands of hectares of rough pasture.

Quarries, mines and collieries are marked on Ordnance Survey maps, and the mineral being worked is usually named. Adjoining buildings may be engaged in processing the raw material, for example brick or tile works may be adjacent to a clay pit, and lime and cement works are often found near chalk or limestone quarries. Where large areas are forested, especially if these take the form of plantations, it is likely that sawmills and woodworking factories may be found.

Many public utilities are easily recognisable on maps. Reservoirs for water supplies can usually be distinguished from lakes by the presence of a dam or embankment and sometimes of water pipelines. Power stations have electricity transmission lines (with pylons), gas works have gasholders. Sewage farms usually occupy low-lying areas near towns.

In urban areas it is sometimes difficult to distinguish between buildings used for commercial, industrial and residential purposes, since their exact use is not always stated. Public buildings such as schools and hospitals are often named and churches are easily recognised.

COMMUNICATIONS

By Water

Our inland waterways, both *rivers* and *canals*, are marked in blue on Ordnance Survey maps. Rivers flow along the bottom of valleys or across plains, and canals often closely follow the rivers in order to avoid large variations in level. Even small variations necessitate the construction of *locks*, in which water is held between lock gates. These open to allow vessels to enter or leave, the level of water being adjusted by means of sluices. The working of locks is shown in figure 2.1. Canals are now less used than formerly. This is due mainly to the slowness of barge traffic and to the fact that many factories are not situated near canals. However, water transport remains the most economical way of transporting bulky, heavy goods, and some

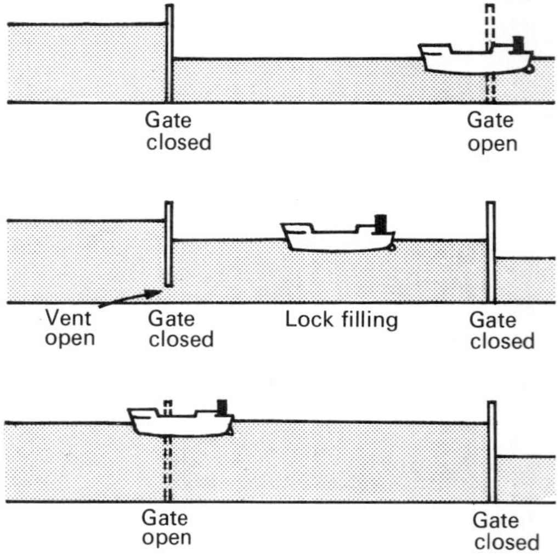

Figure 2.1
A lock—procedure for a vessel going upstream

canals, such as the Aire and Calder Navigation Canal in Yorkshire, and the Manchester Ship Canal, are still of great importance. Rivers and canals are now used by a growing number of pleasure craft.

Many of our largest river mouths and estuaries are navigable by sea-going ships, including large ocean liners and oil tankers, as well as the smaller tramp steamers and coasters, for which they provide safe and sheltered anchorages. Ordnance Survey maps show much detail of harbour installations, such as breakwaters, docks, warehouses, tidal basins and the depth of water.

By Road

Important roads are classified and numbered by the Ministry of Transport, the classification letter and number being printed at intervals along the roads on Ordnance Survey maps. The following types of road are shown:

1 Motorways. These are designed for use by fast passenger and goods vehicles, have dual carriageways and no intersections. They can be identified by the letter M, followed by the number of the motorway.

2 Main Roads. These link our main cities and towns and some have dual carriageways. They are known as A roads. In recent years many by-passes and ring roads have been built so that long-distance travellers do not have to pass through town centres.

3 Secondary Roads. These join both towns and villages and are called B roads.

4 Minor Roads. These carry local traffic and are not classified. Those with a metalled surface are shown on 1 : 50 000 and 1 : 25 000 Ordnance Survey maps in colour. The unmetalled ones are uncoloured and not recommended for motorists.

Roads avoid steep gradients wherever possible. For this reason they may follow valleys into high land, keeping just above the flood level of the river and making use of gaps or cols, if these are present. Deep, narrow valleys are usually avoided and high land may be climbed by a series of hairpin bends, the contours being crossed diagonally. On 1 : 50 000 Ordnance Survey maps a steep gradient on a road is indicated by an arrow head (for gradients between 1 in 7 and 1 in 5) or by two arrow heads (for a gradient of 1 in 5 or steeper).

When crossing lowland, roads avoid marshes and areas liable to flooding. Rivers are usually crossed where the marsh is narrowest, the bridge being approached on low embankments. Ancient bridges may occupy the sites of former fords, or make use of islands.

Railways

Railways are generally as straight as possible since trains cannot negotiate sharp bends. Gradients seldom exceed 1 in 60. In order to maintain as even a gradient as possible valleys are often crossed on embankments or viaducts. When passing through high land railways usually follow the valleys, crossing and recrossing the river if it winds and making use of cuttings and tunnels where necessary.

Railways provide a speedy and reliable means of transport for passengers, mail and goods. This is especially so with heavy goods such as coal and machinery, and perishable ones like milk and fresh vegetables. Special sidings have been built to serve collieries, factories and docks.

TOURISM

Tourist information enables map users to locate main attractions in an area and to assess the importance of tourism.

Exercises

1 Give an example from your own locality of (*a*) a mineral working, stating its location, the type of mineral and its uses, (*b*) a farm, stating its location, name and the type of farming, and (*c*) a factory, stating its location, name and the goods produced, and (*d*) a tourist attraction, stating its location, name and type of attraction.

2 Draw a sketch map of any town you know, showing the rivers and high land (if any), the main roads and railways, the built-up areas, the principal industries and any local public services such as waterworks, power stations, gasworks and sewage disposal works.

EXERCISES IN MAP READING

The following sixteen pages give further practice in map reading. For reference purposes, the standard Ordnance Survey symbols are reproduced on pages 12 and 13.

1:50 000 Second Series Map
CONVENTIONAL SIGNS

Ordnance Survey

ROADS AND PATHS **Not necessarily rights of way**

Junction number · Service area · M 1 · 3 · Elevated — Motorway (dual carriageway)

M 3 — Motorway under construction

Unfenced · Footbridge · A 40 (T) — Trunk road

Dual carriageway — Main road

Main road under construction

B 284 — Secondary road

A 855 · B 885 — Narrow road with passing places

Bridge — Road generally more than 4m wide

Road generally less than 4m wide

Other road, drive or track

Path

Gradient : 1 in 5 and steeper   1 in 7 to 1 in 5

Gates   Road tunnel

Ferry P — Ferry (passenger)   Ferry V — Ferry (vehicle)

RAILWAYS

Track multiple or single

Track narrow gauge

Freight line, siding or tramway

Station (a) principal (b) closed to passengers

LC — Level crossing

Embankment

Cutting

Bridges, Footbridge

Tunnel

Viaduct

ROCK FEATURES

PUBLIC RIGHTS OF WAY (Not applicable to Scotland)

Footpath
Bridleway
} Public paths

Road used as a public path

Byway open to all traffic

HEIGHTS

When standard metric contours are not available the contour interval is 50 ft with values shown to the nearest metre

50 — Contours are at 10 metres vertical interval

· 144 — Heights are to the nearest metre above mean sea level

Heights shown close to a triangulation pillar refer to the station height at ground level and not necessarily to the summit.

1 metre = 3·2808 feet   15·24 metres = 50 feet

Public rights of way indicated by these symbols have been derived from Definitive Maps as amended by later enactments or instruments held by Ordnance Survey on (date) and are shown subject to the limitations imposed by the scale of mapping

The representation on this map of any other road, track or path is no evidence of the existence of a right of way.

Danger Area — MOD Ranges in the area. Danger! Observe warning notices

WATER FEATURES

- Marsh or salting
- Lake
- Canal, Lock and towpath
- Canal (dry)
- Aqueduct
- Footbridge
- Normal tidal limit
- Lighthouse(in use and disused)
- Beacon

- Slopes
- Cliff
- Flat rock
- Low water mark
- Mud
- High water mark
- Sand
- Dunes
- Shingle

Conventional signs for 1 : 25 000 maps

GENERAL FEATURES

- Electricity transmission line (with pylons spaced conventionally)
- Pipe line (arrow indicates direction of flow)
- Buildings (ruin)
- Public buildings (selected)
- Quarry
- Spoil heap, refuse tip or dump
- Coniferous wood
- Non-coniferous wood
- Mixed wood
- Orchard

- Radio or TV mast
- Church or Chapel: with tower / with spire / without tower or spire
- Chimney or tower
- Glasshouse
- Graticule intersections at 5′ intervals
- Heliport
- Triangulation pillar
- Windmill with or without sails
- Windpump
- Park or ornamental grounds

ABBREVIATIONS

P	Post office
PH	Public house
MS ·	Milestone
MP ·	Milepost
CH	Clubhouse
PC	Public convenience (in rural areas)
TH	Town Hall, Guildhall or equivalent
CG	Coastguard

ANTIQUITIES

VILLA	Roman
Castle	Non-Roman
⚔	Battlefield (with date)
☆	Tumulus (Ancient mound)
+	Position of antiquity which cannot be drawn to scale
𝔪	Ancient Monuments and Historic Buildings in the care of the Secretaries of State for the Environment, for Scotland and for Wales and that are open to the public

The revision date of archaeological information varies over the sheet

BOUNDARIES

- National
- London Borough
- District
- County, Region or Islands Area
- National Park or Forest Park
- NT — National Trust: always open / opening restricted
- NTS (in red or blue) National Trust for Scotland
- FC — Forestry Commission
- Pedestrians only - observe local signs

TOURIST INFORMATION

- Information Centre
- Selected places of tourist interest
- Viewpoint
- Parking
- Picnic site
- Camp site
- Caravan site
- Youth hostel
- Golf course or links
- Bus or coach station
- Public telephone
- Motoring organisation telephone
- Public convenience (in rural areas)

Note Outdoor Leisure maps have additional signs.

Made and published by the Ordnance Survey, Southampton

Exercise A (map extract 1)

1 Give six-figure references for (*a*) a windmill close to a church with a tower, (*b*) a radio or TV mast, (*c*) a triangulation pillar, (*d*) flat rocks at the end of a narrow promontory.

2 Identify and locate three different features of archaeological interest. What is each of the features you refer to?

3 Name (*a*) THREE natural attractions, and (*b*) THREE man-made attractions which help to explain why Weston-super-Mare is a well-known holiday resort.

4 Quoting map evidence, list the methods of travel available to a visitor to Weston-super-Mare who does not own a car.

5 If you were given the opportunity to have a house built for yourself at one of the following four sites, which would you choose? Why did you not choose one of the others? (*a*) 377625, (*b*) 305584, (*c*) 320590, (*d*) 363570.

6 Study the photograph on this page and then answer the following questions:
(a) in which direction was the camera pointing?
(b) name three things which the photograph shows but not the map;
(c) name three things which the map shows but not the photograph.

Exercise B (map extract 1)

Draw a simplified sketch map, on half the scale of the extract, to show the relief of the whole area, marking in the 10 m and 100 m contours, the coastline and the main river.

Then, using the map which you have made as a guide, write a brief account of the relief of the area.

WESTON-SUPER-MARE (*Aerofilms*)

The Highlands are made of ancient igneous and metamorphic rocks which give rise to a very rugged scenery. The relief has also been greatly affected by glaciation. Valleys were overdeepened by ice, and those along the coast were flooded by the post-glacial rise in sea level. These flooded valleys are known as fiords in Norway and lochs in Scotland, and often have deeper water within them than at their entrances. The glaciated valleys inland have flat floors and steep sides, and contain lakes, some of which are so deep that their floors are below sea level.

Exercise A (map extract 2)

1 What conventional signs are shown at (*a*) 817266, (*b*) 830286 to 834283, (*c*) 853261 to 837276, (*d*) 888298?

2 In what ways does the coast at 872247 differ from that at 815245?

3 What relief features are to be found at (*a*) 880236, (*b*) 838258, (*c*) 880250, (*d*) 845285?

4 Locate by grid references three areas of woodland. Between what heights do each of these extend? What conclusions can you come to about the distribution of woodland in the area?

5 Study the course of the A816(T) from 840232 to Oban Quay. (*a*) What is the distance along the road between these two points? (*b*) What is the highest point reached by the road? Give its grid reference. (*c*) Explain why the road follows the course it takes.

6 Name the occupations for which there is evidence on the map. Give the evidence with a reference.

7 Which portion of the photograph appears on the map extract? Describe the site of Oban and the services it provides for the surrounding area, using information both from the map and the photograph.

Exercise B (map extract 2)

Draw a sketch map of the area between eastings 81 and 88 and northings 24 and 27. Draw contours at 100, 200 and 300 metres. Shade the areas which are uncovered at low tide, and show the rocks along the coast.

OBAN AND A PORTION OF THE SCOTTISH HIGHLANDS (*Aerofilms*)

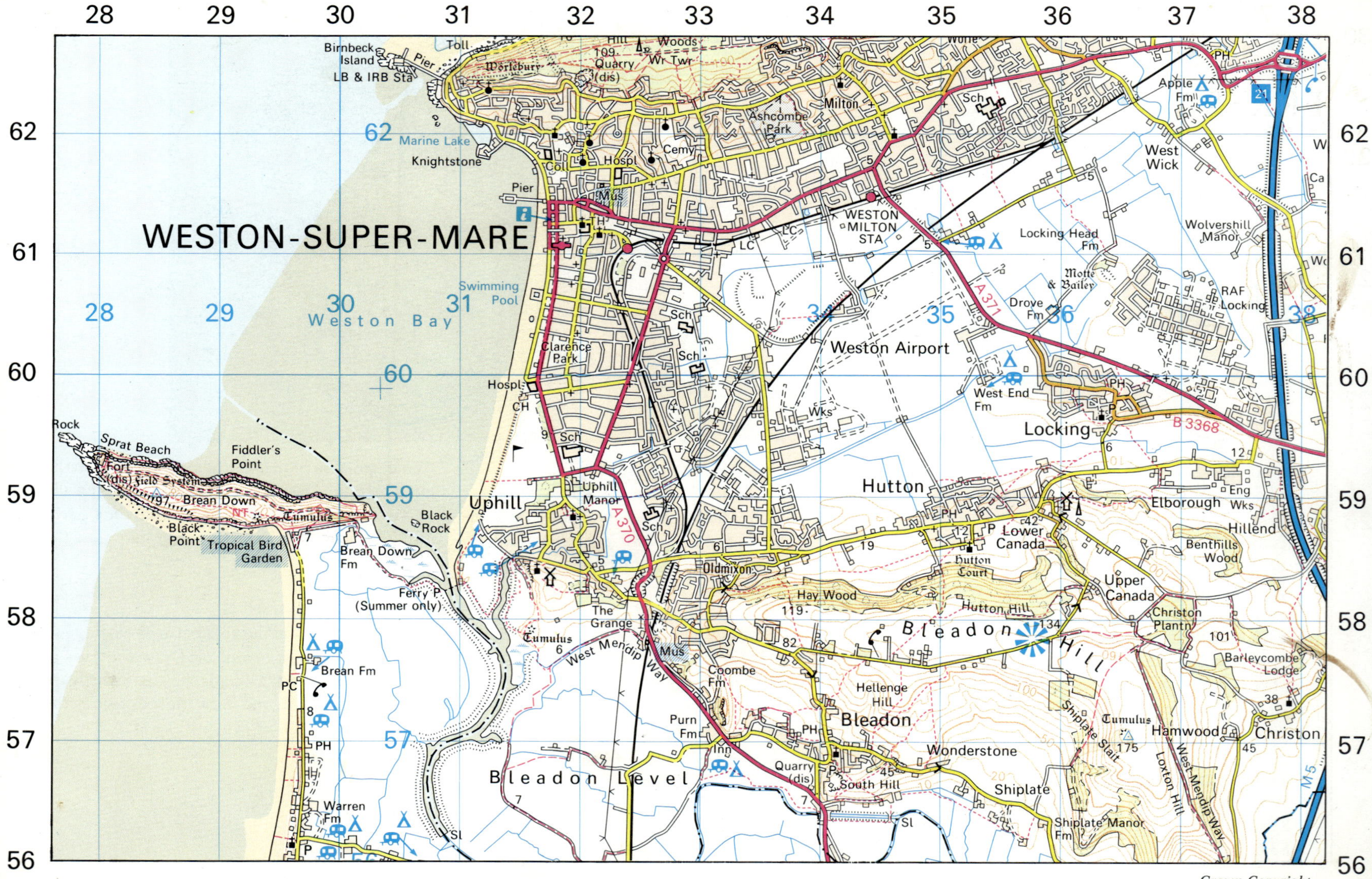

 Map extract 1 **WESTON-SUPER-MARE** 1 : 50 000

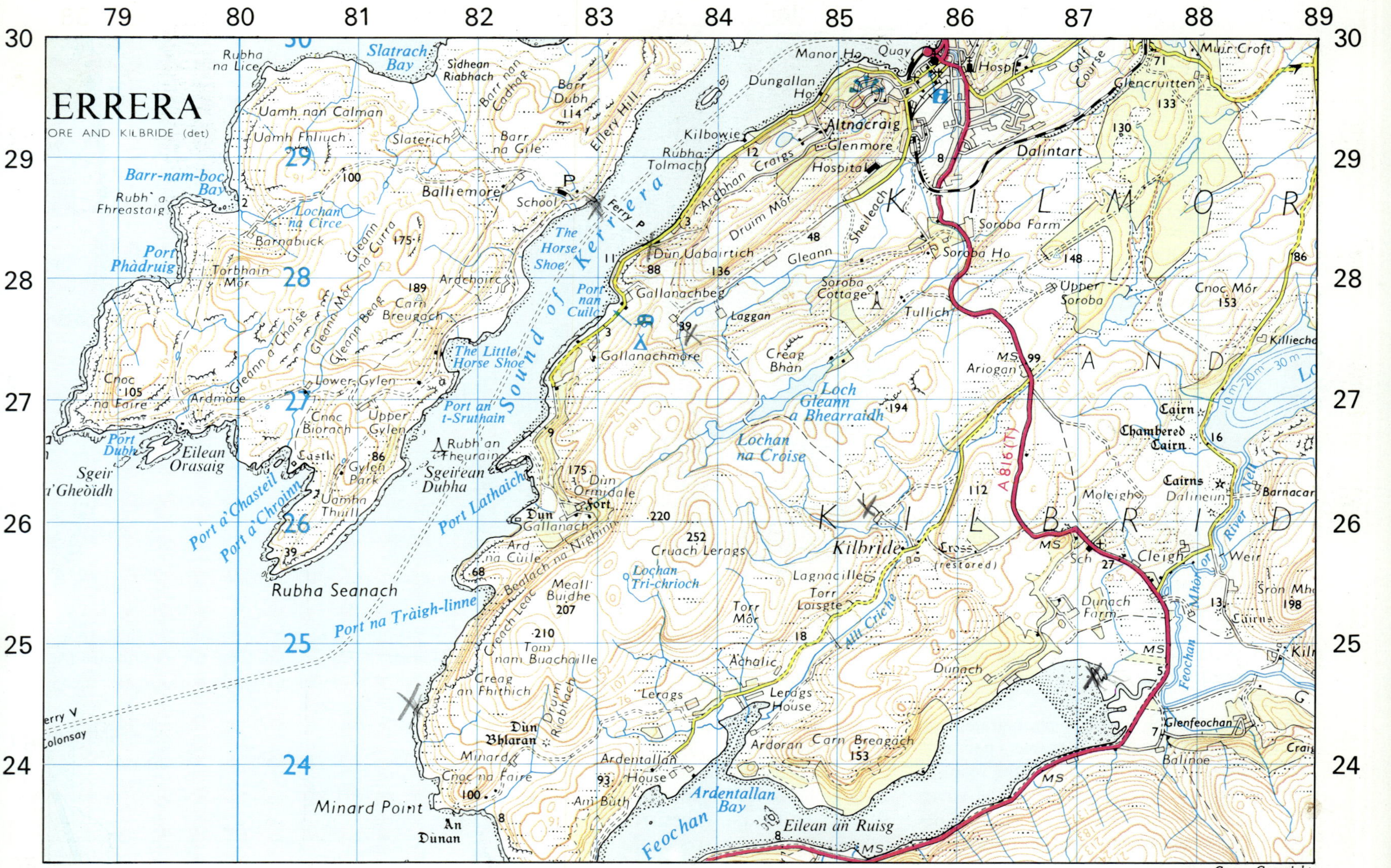

OBAN AND A PORTION OF THE SCOTTISH HIGHLANDS 1 : 63 360 **Map extract 2**

The geological structure of the Malham area is shown in the figure on this page. Because of faulting, Silurian shales are at the surface around Malham Tarn, Carboniferous Limestone in the centre and Yoredale beds in the south. The shales and Yoredale beds are both impermeable while the limestone is pervious, which means that rain water passes underground down the joints in the rock. Two streams pass underground where they reach the limestone at 882659 and 894656. The stream that reappears at Malham Cove is the one that passed underground at 882659.

Exercise A (map extract 3)

1 State the meaning of the following abbreviations used on this map, and give a six-figure reference for each: F.B., B.S., Spr.
2 What type of vegetation is to be found at (*a*) 907663, (*b*) 908658, (*c*) 898640?
3 Describe the relief features at (*a*) 911660, (*b*) 920643, (*c*) 913637.
4 Give directions to a walker at Street Gate crossroads (905656) who wishes to reach Gordale House (915635) by paths and roadways shown on the map. He has a compass but has lost his map.
5 What evidence is there of former mining in the area?
6 Calculate the average gradient along the road from Gordale Bridge (913635) to the T-junction at 923635.
7 Identify the scene on the photograph. (*a*) Towards which direction is the camera pointing? (*b*) What is the map reference of the roadside farmhouse? (*c*) Make a sketch of the scene, labelling any distinctive features, and then describe the landforms you have illustrated.

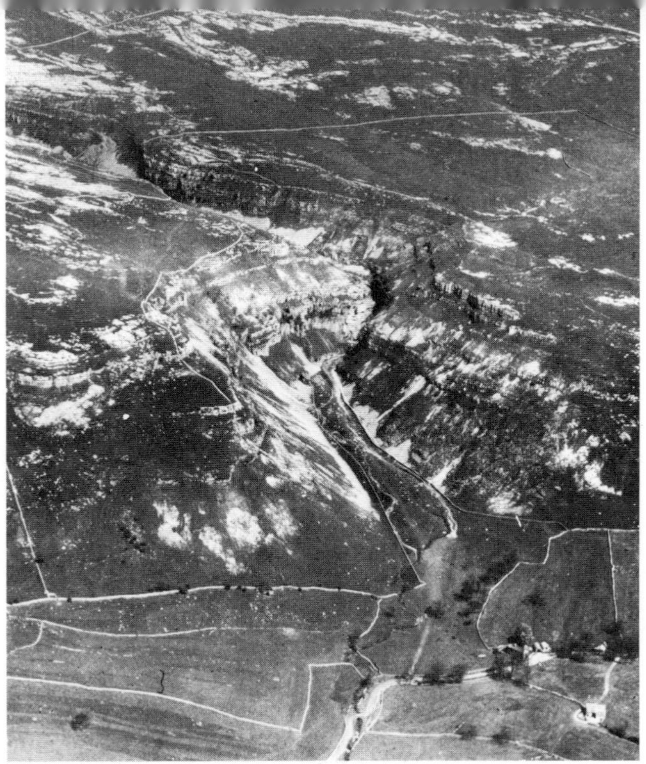

Exercise B (map extract 3)

Draw a simplified sketch map of the area to the east of easting 89 on the same scale as the map extract. Name Malham Cove as well as the streams, and print SWALLOW HOLE and GORGE in the correct places. Show, with a broken line, the scarp edge which runs from 920635 westwards to Malham Cove, and then south-west to the edge of your map. In large block letters print the words SILURIAN SHALES, LIMESTONE PAVEMENT and YOREDALE BEDS in the correct places on the map.

What kind of rock is Carboniferous Limestone and what indications are there for its presence in this area? Use evidence from both the map extract and the photograph.

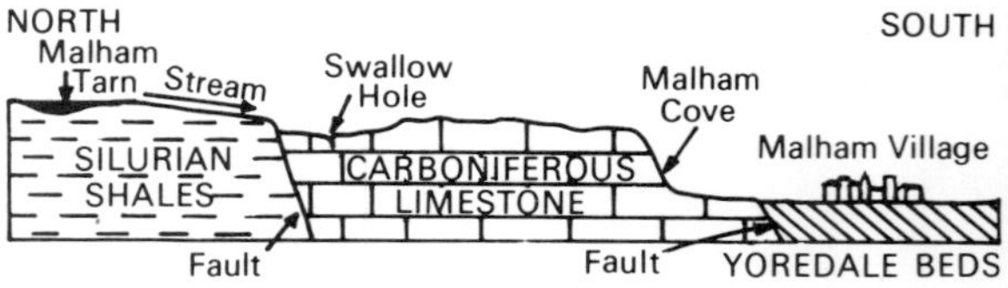

THE MALHAM AREA (*Aerofilms*)

This part of Norfolk is undulating chalk country, with a low-lying coastal plain in the north. Large parts of the chalk are covered with a light boulder clay, which was left behind when the North Sea ice sheet retreated at the end of the Ice Age. The shore here consists of sand and shingle, with a detached spit known as Scolt Head Island. Prevailing waves from the north-east have caused an east–west longshore drift and the same waves have thrown up the shingle to form the spit. Other waves from the north-west have turned the end of the spit to form laterals. In the shelter of the shingle banks, extensive salt marshes have formed, while in places, sand from the shore has been blown inland to form dunes. Much of this sand has been anchored by the planting of marram grass and belts of pine trees.

Exercise A (map extract 4)

1 a) Express the scale used in centimetres to the kilometre.
b) What is the vertical interval used?

2 How far is it by the main road, A 149, from the church at Burnham Deepdale to the crossroads at Holkham (892440)? Give your answer to the nearest kilometre.

3 a) Identify the vegetation type at (i) 796441, (ii) 800449, (iii) 800432.
b) Identify and give a six-figure reference for each of these abbreviations: MS, PH, P, NT.

4 In what ways do the site and position of Burnham Overy Staithe differ from those of Burnham Thorpe?

5 Holkham Park has been landscaped in the past at very considerable expense. What different features on the map provide evidence of this?

6 Pair the following grid references with the descriptions given: (*a*) 862457, (*b*) 818410, (*c*) 857406, (*d*) 849450, (*e*) 8744.
(i) dry valley, (ii) shallow valley with stream, (iii) coastal sand dunes, (iv) coast plain, (v) estuary.

Exercise B (map extract 4)

Draw a map on the scale 1:25 000 for the area north of northing 43 and west of easting 83. Indicate on it the areas of marsh, sand dunes, shingle, areas exposed at low water mark and former marshland which has been drained.

Then write a description of the scenery of the area covered by your map, using the evidence both on the extract and the photograph. How can you tell that some of the former marsh has been drained?

A COASTAL AREA OF NORFOLK (*Aerofilms*)

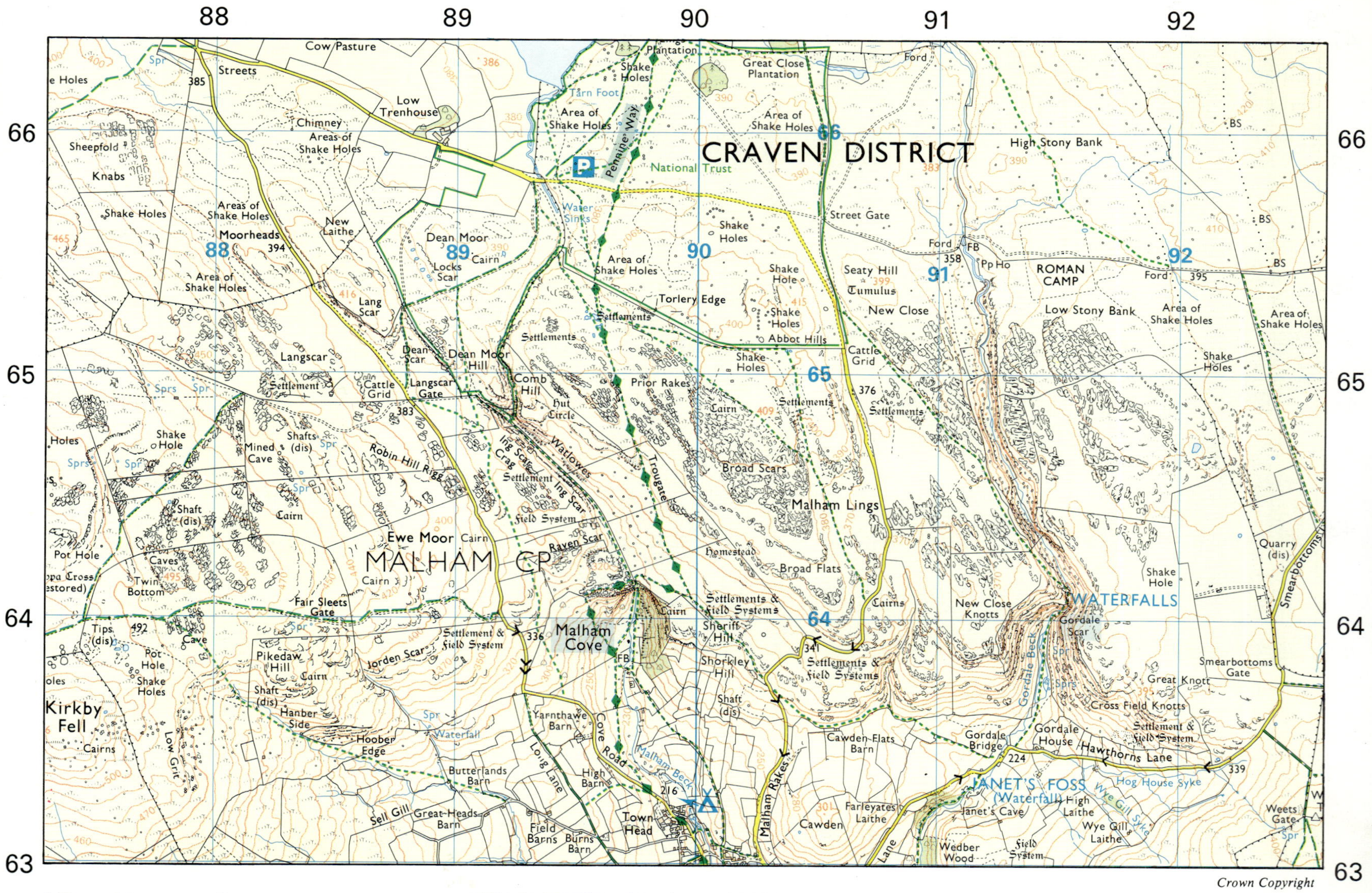

20] **Map extract 3** **THE MALHAM AREA** 1 : 25 000

Crown Copyright

A COASTAL AREA OF NORFOLK 1 : 50 000

Map extract 4

The Thames basin downstream from Tower Bridge is a densely populated area which contained most of the main London docks. They are now closed and most traffic goes through Tilbury Docks, very much further downstream. The docks were excavated in the soft London Clay and alluvium which form most of the lower land. To the north of the docks is London's East End, where slum clearance has provided space for the construction of many tower blocks of flats. The riverside is crammed with old factories and warehouses as far downstream as Woolwich. There are also power stations using coal or oil, also brought by water. To protect London from flooding, the Thames Barrier has been built on this stretch of the river. The London Docklands Development Corporation was set up in 1981 to promote the economic redevelopment of the docks areas. Amongst the proposed developments are an airport for short take-off and landing aircraft at the Royal Docks, a new light railway and housing, office and industrial developments.

View north-east from G. R. 340800. St Katherine's Docks can be seen in the foreground

THE EAST END OF LONDON (*Aerofilms*)

Exercise A (map extract 5)

1 Draw a linear scale for this map, showing distances in kilometres, with one kilometre subdivided into hundreds of metres.

2 How far is it from the road junction at 403832 to the junction for Woolwich Ferry at 433799 (*a*) in a straight line, (*b*) by road, in kilometres and metres? Use the linear scale you have just made.

3 What is the direction of the Woolwich Ferry from the road junction at 403832 (*a*) to the nearest compass point, (*b*) as a three-figure bearing?

4 Identify (*a*) the tourist attraction at 383779, (*b*) the feature at 370808.

5 Which are the two ways vehicles can cross the Thames along this stretch of the river? Give a grid reference for each.

6 How does the relief of the land to the north of the Thames differ from that to the south?

7 Give directions to a driver who has no map, is outside the station at 438788, and must get to the hospital at 351793.

Exercise B (map extract 5)

1 What bridge can just be seen in the bottom left of the photograph?

2 What historic building lies west of St Katherine's Dock?

3 St Katherine's Dock has been closed to dock trade for many years. Using evidence on the map and the photograph, describe the present-day use of the dock and the area immediately adjacent.

4 What does the photograph show that was mentioned in the description of the area but which cannot be identified from the map?

Exercise A (map extract 6)

1 Give the names of four different paths leading to the summit of Snowdon.
2 Put the following lakes in order of the height of their surfaces above sea level (the highest first): Llyn Gwynant, Glaslyn, Llyn Llydaw, and Llyn Du'r Arddu.
3 Pair the following references and descriptions: (*a*) 565564, (*b*) 625532, (*c*) 576566, (*d*) 6553; (i) a ridge with very steep sides, (ii) a broad valley with flat floor and steep sides, (iii) a col, (iv) a V-shaped valley.
4 Could you see a climber who is at 643578 when you are at the milestone at 623570? Give reasons for your answer with a sketch section.
5 Give specific map evidence to support the view that, while many long established means of employment in this area have disappeared, catering for leisure activities in nearby villages is now a major activity.
6 Study the photograph and then state (*a*) the lakes shown, (*b*) in which direction the camera was pointing, (*c*) the heights of the two summits, (*d*) what man-made features can be recognised.

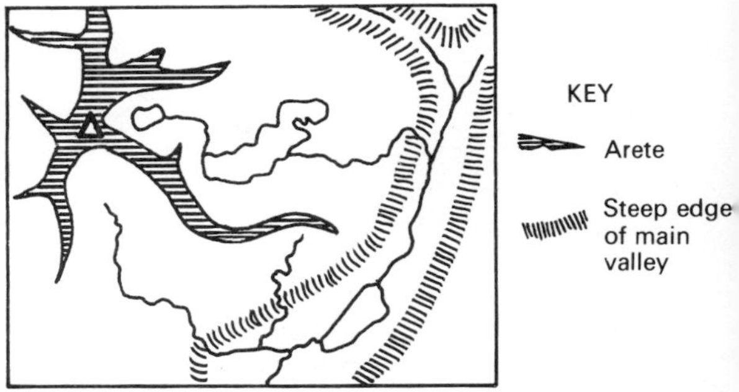

Exercise B (map extract 6)

(to be undertaken only after studying Chapter 9) On this page is a simplified map of the area between eastings 60 and 66 and northings 51 and 56. Make a copy of this map, naming the lakes, streams, arêtes and the Snowdon summit.

Then make a list of the features shown on the map which point to the area having been glaciated during the Ice Age, with a brief explanation of each.

THE SNOWDON AREA (*Aerofilms*)

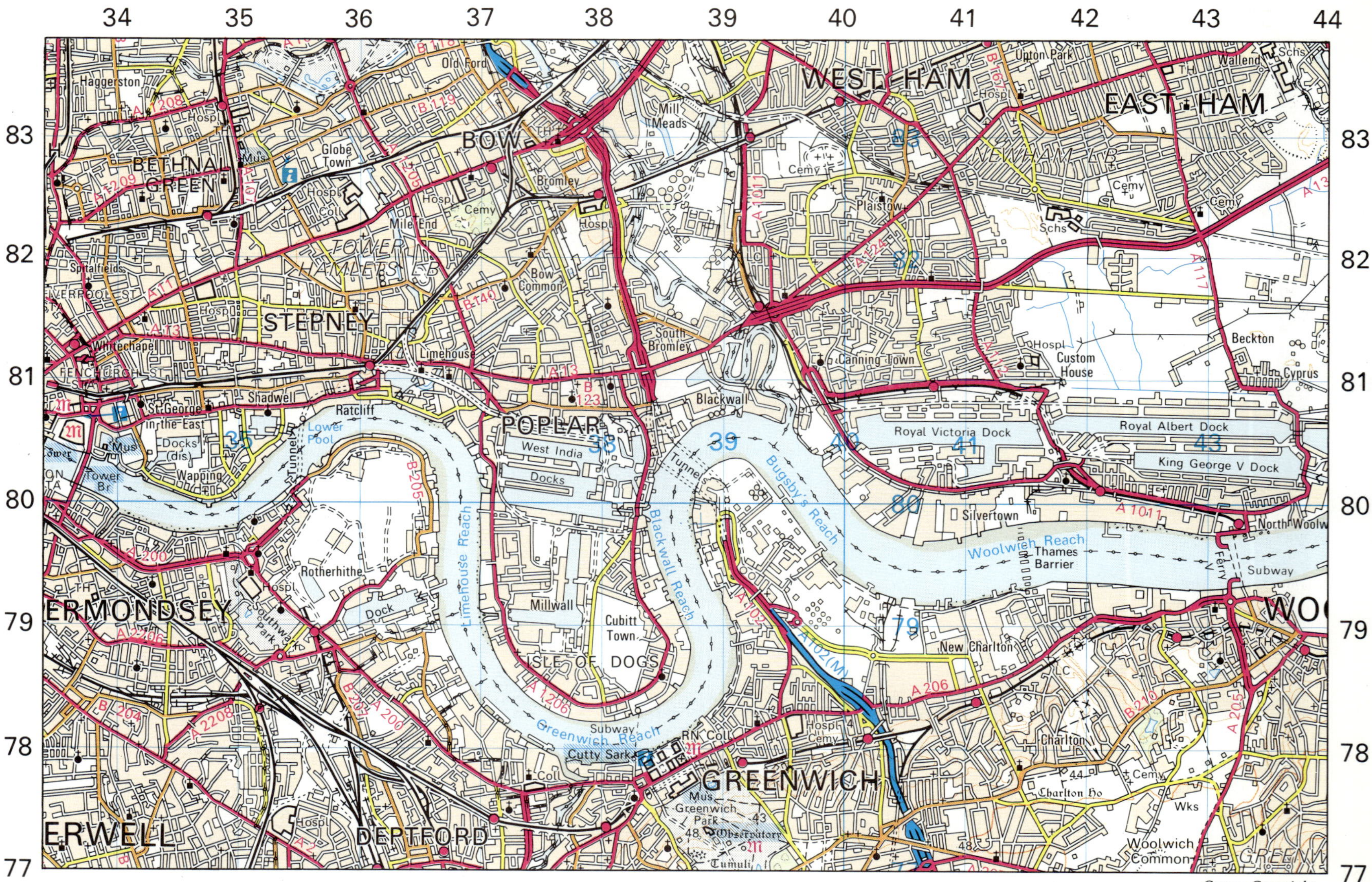

 Map extract 5 **THE EAST END OF LONDON** 1 : 50 000

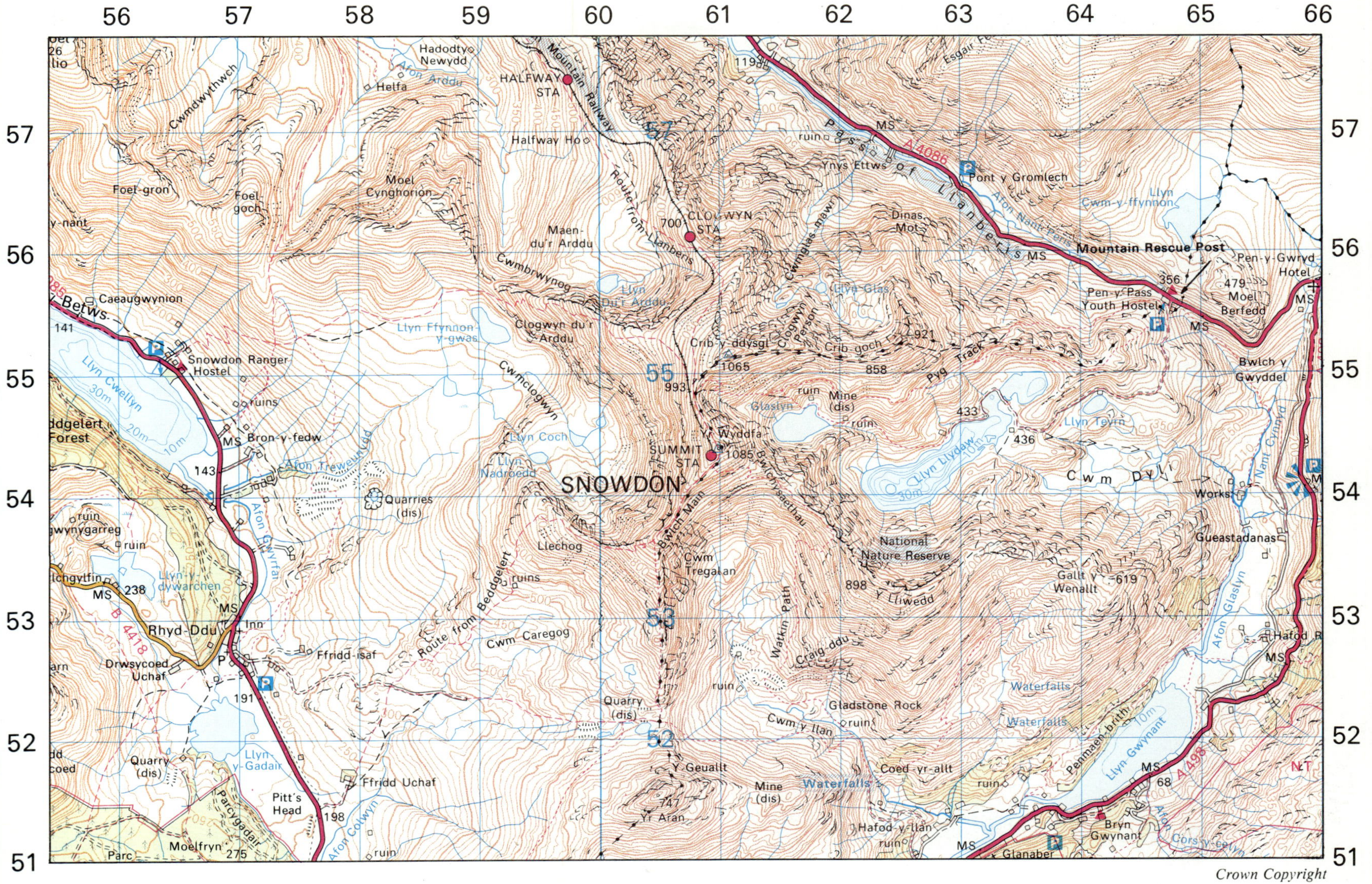

THE SNOWDON AREA 1 : 50 000

Map extract 6

The southern half of the map covers part of the Lambourn Downs, which are chalk uplands, having a north-facing scarp slope and a dip slope to the south. Some details of chalk country are given on page 81. North of the chalk scarp is a narrow belt of Upper Greensand, and then a broad expanse of Gault Clay. The clay is a soft, impermeable rock which has been worn down to form a low-lying vale. Streams start at the foot of the scarp, and the villages which have grown up around these springs are of the type known as 'spring-line villages'. There is much evidence of prehistoric life on the Downs, including Uffington Castle, an Iron Age hill fort.

Simplified geological section across the area

Exercise A (map extract 7)

1 What is the scale of the map in centimetres to the kilometre? What vertical interval is used and what area does each grid square represent?

2 Give the six-figure grid references for (*a*) a telephone close to an old mill pond, (*b*) a disused quarry, (*c*) a post office in a village.

3 Which of the following land forms—plain, valley, spur, concave slope, convex slope—are shown at 297873, 301868, 305868, 285885, 300873?

4 Locate by means of grid references the highest and lowest points indicated on the map, and give their heights.

5 Draw an accurate cross section from the top or northern edge to the bottom or southern edge of the map along easting 30, using a vertical scale of 50 metres to a centimetre. Indicate on the section the positions of Stockholm Farm and Uffington Castle. Can the one be seen from the other?

Exercise B (map extract 7)

Draw a rectangle to represent the area shown on the map, but on half the scale. Insert the contours for 120 and 210 metres, and all the streams. Mark with a dot and name the villages. Print CHALK UPLAND, SCARP SLOPE and CLAY VALE in the correct place.

What does the map tell you about the relief, drainage and settlements of (*a*) the Chalk Upland and (*b*) the Clay Vale? What additional information does the photograph give?

A RURAL AREA IN OXFORDSHIRE (*Aerofilms*)

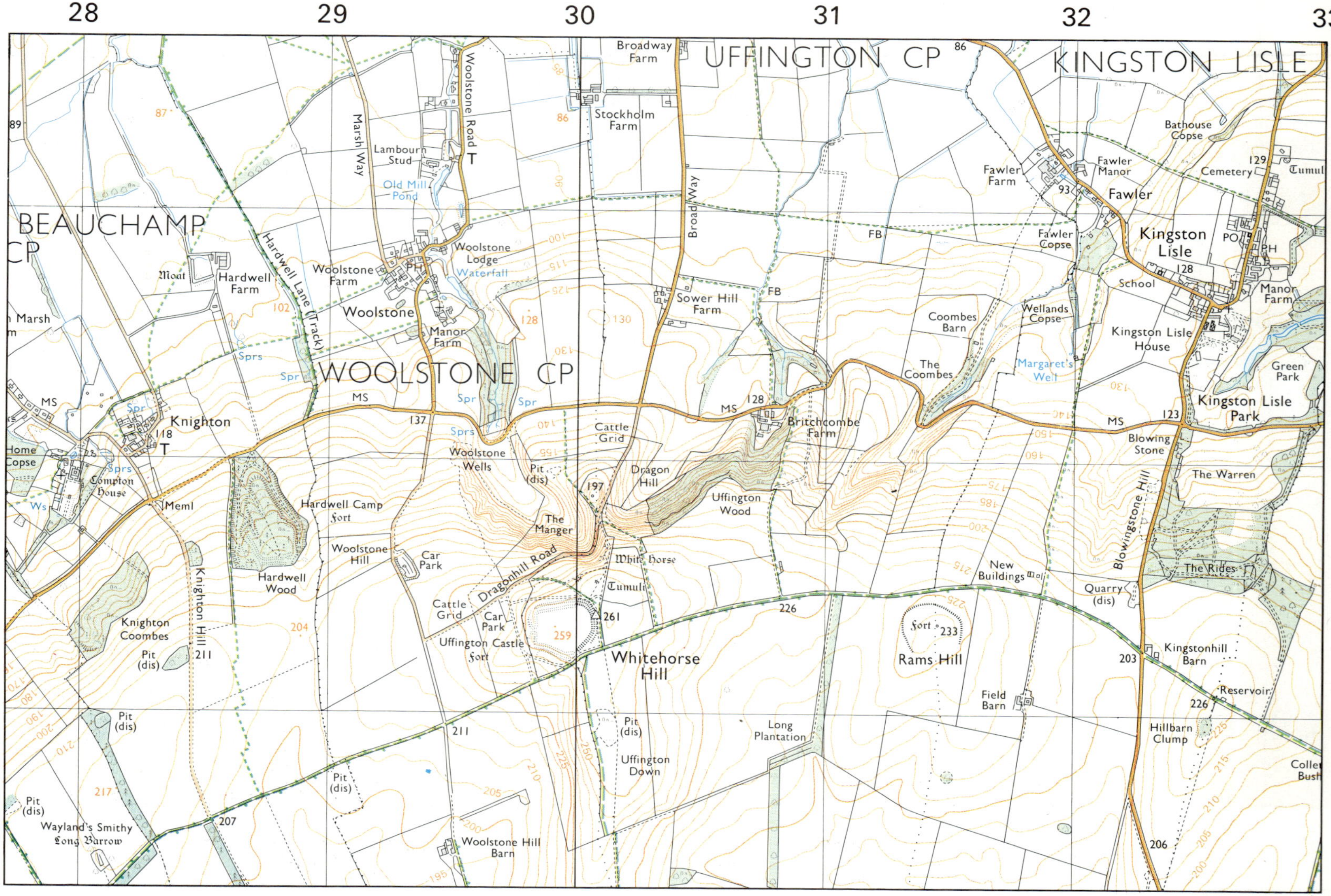

 Map extract 7 **A RURAL AREA IN OXFORDSHIRE** 1:25 000

PART TWO

The Earth and its Atmosphere

Chapter 3

The Earth and its Movements

The Solar System

The tiny part of the Universe to which we belong is called the Solar System and consists of the Sun and the various bodies that move round it. The Sun itself is a star, with a diameter of about 109 times that of the Earth. Physical reactions within it produce enormously high temperatures and cause heat and light energy to be radiated into space from its gaseous surface.

The most important of the bodies that move round the Sun are called *planets*. They follow elliptical paths, known as *orbits*, at varying distances from the Sun, and are held in their orbits by the Sun's gravitational attraction. Their names and relative positions are shown in figure 3.1, and it will be seen that several of them have moons. Neither the planets nor their moons give out light of their own, but they all shine because they reflect the light of the Sun.

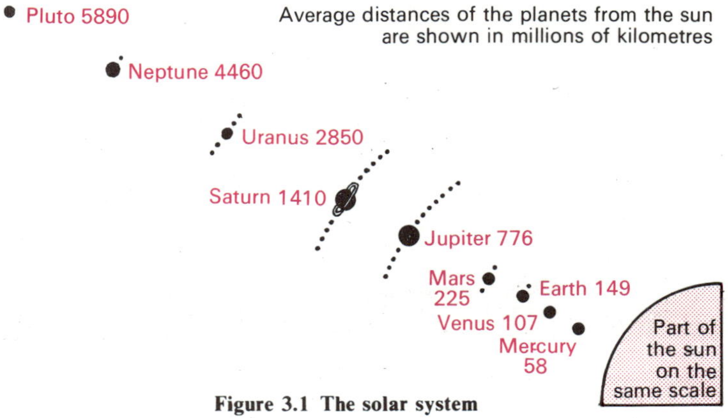

Figure 3.1 The solar system

The Earth and the Moon

In shape the Earth is nearly a sphere, having a slight bulge at the Equator. Its diameter is just under 13 000 kilometres and its circumference about 40 000 kilometres. It moves in two different ways:

1 It rotates or spins round its own axis once in 24 hours, or one day. The ends of the axis are known as the North and South Poles.

2 It revolves in orbit round the Sun once in 365¼ days, or one year. To keep the calendar correct a leap year has been arranged every fourth year, in which an extra day (February 29) is added. The mean distance of the Earth from the Sun is about 150 million kilometres, and its velocity in orbit varies according to its position in the ellipse, being greatest when it is closest to the Sun and least when it is farthest away.

The Moon is without atmosphere or water, and presumably without life as we know it. It revolves round the Earth in about 28 days, or one lunar month, at a mean distance of 380 000 kilometres, and, as a result of the fact that it also rotates on its own axis once during each revolution round the Earth, always presents the same face to the Earth. Only half of the Moon is illuminated at a time by the Sun, and from the Earth varying portions of the illuminated half are visible on different days, depending on the Moon's position in its orbit. These *phases* of the Moon are shown in figure 3.2.

The Moon is of great importance to us because its gravitational attraction, together with that of the Sun, is responsible for the tides. More will be said of this in Chapter 11.

Figure 3.2 The phases of the Moon

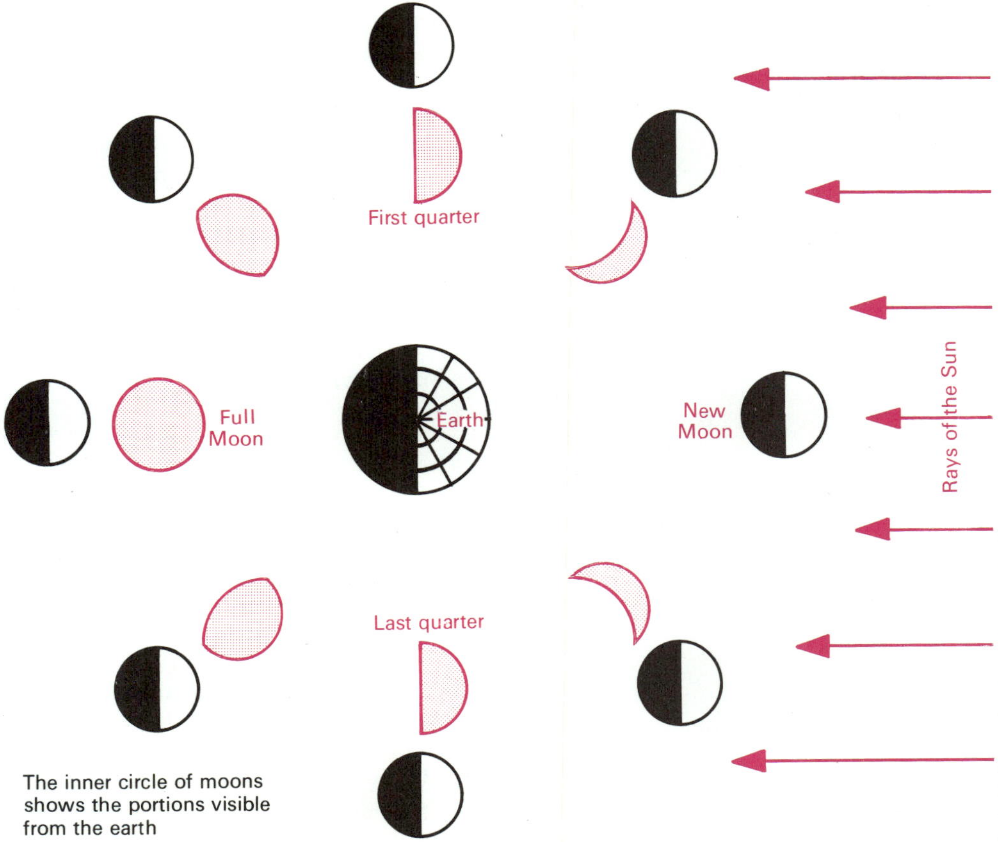

The Effects of the Earth's Movements

The direction of the Earth's rotation is from west to east. The Sun therefore appears to rise in the east, reaches its greatest altitude[1] at noon, and sets in the west, giving rise to the sequence of day and night. night.

Figure 3.3 The altitude of the Sun

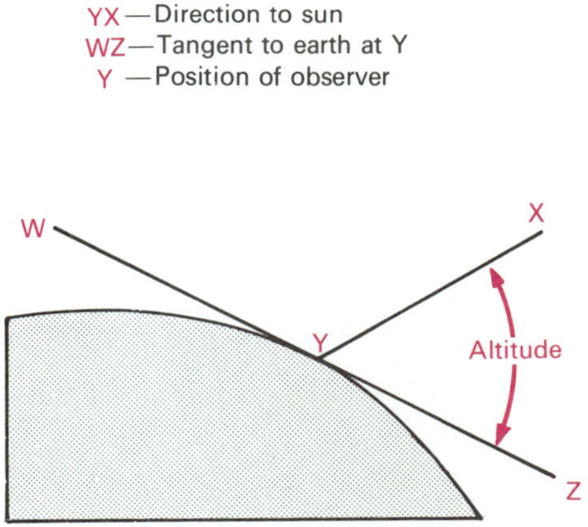

The Earth's revolution round the Sun has a more complicated effect, producing the seasons and variations in the length of day and night. This is the result of the fact that the Earth's axis is inclined at an angle of $66\frac{1}{2}°$ to the plane made by its orbit round the Sun. This may be seen in figure 3.4, which

[1] When applied to a heavenly body altitude means the angle of elevation above the horizon, measured from an observer. In figure 3.3 the angle XYZ is the altitude of the Sun, measured from an observer at Y. This angle can be obtained by using a sextant or clinometer.

shows the Earth in four positions, corresponding to each of the four seasons. It should be noted that:
On June 21, the northern summer, the North Pole is tilted towards the Sun and the South Pole away from it. Owing to this (i) the Sun is vertically overhead at the Tropic of Cancer and (ii) whilst the Equator has 12 hours day and 12 hours night, the length of daylight increases northwards until within the Arctic Circle it lasts the full 24 hours, and decreases southwards until within the Antarctic Circle there is no daylight at all.
On December 21, the northern winter, the North Pole is tilted away from the Sun and the South Pole towards it. Owing to this (i) the Sun is vertically overhead at the Tropic of Capricorn and (ii) whilst the Equator has 12 hours day and 12 hours night, the length of daylight decreases northwards until within the Arctic Circle there is no daylight at all, and increases southwards until within the Antarctic Circle it lasts the full 24 hours.
On March 21 and September 22, the northern spring and autumn, the North and South Poles are equidistant from the Sun, so that (i) the Sun is vertically overhead at the Equator and (ii) everywhere on the Earth has 12 hours day and 12 hours night.

March 21 and September 22 are often referred to as the *equinoxes* (from the Latin, equal night), while June 21 and December 21, the dates on which the overhead Sun reaches its farthest point from the Equator, are called the *solstices* (from the Latin, sun stops).

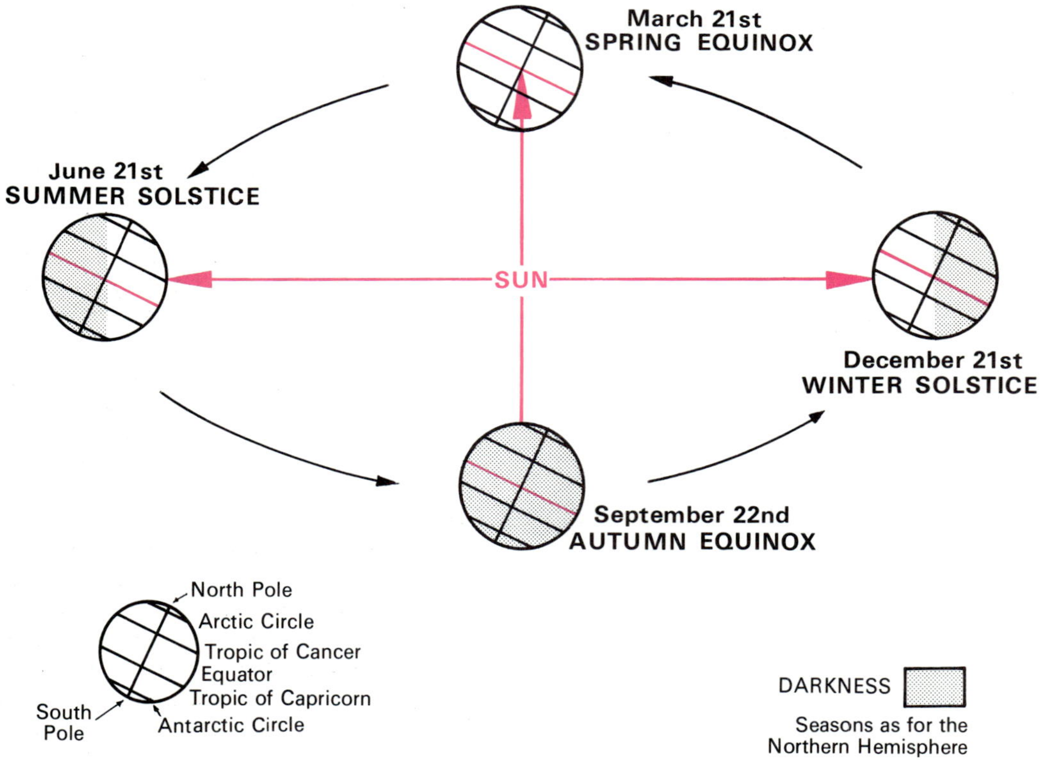

Figure 3.4 The causes of the seasons

LATITUDE AND LONGITUDE

Reference has already been made in Chapter 1 to the degrees of latitude and longitude, whereby any point on the Earth's surface can be identified. Navigators of ships and aircraft are able to find their latitude and longitude by measuring, with a sextant and compass, the altitude and bearing of the Sun or certain stars, and looking up these readings in the Nautical Almanac, against the correct date and time.

Latitude

This is the angle north or south of the Equator, measured from the centre of the Earth along the same line of longitude. In figure 3.5 A is a place on the Earth's surface, O is the centre of the Earth and B is a point on the Equator on the same line of longitude as A. The angle BOA is the latitude of A north of the Equator.

Lines of latitude form parallel circles on the globe and include the Equator itself (0°), the Tropics of Cancer and Capricorn ($23\frac{1}{2}$° north and south) and the Arctic and Antarctic Circles ($66\frac{1}{2}$° north and south). The North and South Poles are at 90° north and south.

One degree of latitude, measured along a line of longitude, is about 111 kilometres. It is useful to remember this when estimating the length of a country in a north–south direction, from a map on which latitudes are marked.

Longitude

This is the angle east or west of the *Greenwich Meridian* (often called the *Prime Meridian*), measured from the centre of the Earth along the Equator. In figure 3.5 the meridian which passes through A crosses the Equator at B, whilst the Greenwich Meridian crosses it at C. The angle COB is the longitude of A, west of Greenwich.

The Greenwich Meridian is taken as 0° and the longitude increases both eastwards and westwards to 180°, giving a total of 360°.

Longitude and Time

All places on the same meridian have noon at the same moment. Since the Earth rotates from west to east places to the east are ahead in time and places to the west are behind. The Earth rotates through 360° of longitude in 24 hours, so that it takes one hour to pass through 15°. Thus we can find the difference between the local times of two places by dividing the difference in longitude by 15, remembering that places to the east have time in advance of those to the west.

It would be confusing if within a country each place kept its own local time. Small countries such as the United Kingdom maintain one *standard time* throughout, whilst large countries such as the USSR, USA and Australia are divided into *time zones*. The standard time in the United Kingdom is known as *Greenwich Mean Time*, which is the local time at Greenwich Observatory, London.

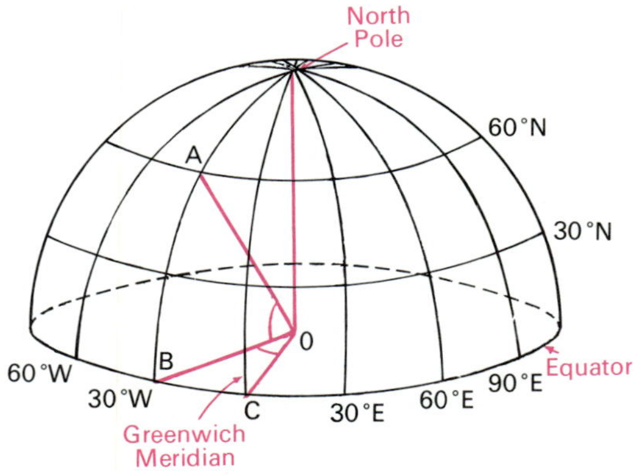

Figure 3.5
Latitude and longitude—for the northern hemisphere

∠ BOC—Longitude 30 °W of Greenwich Meridian
∠ BOA—Latitude 60 °N of Equator

The International Date Line

If a ship sails round the world in an easterly direction its clocks are advanced by 1 hour for every 15° of longitude. On return to port it will have steamed through 360° of longitude and its clocks will have been advanced by 24 hours. The ship's calendar would therefore show the date to be one day ahead of the date ashore. On the other hand, if the ship steams round the world in a westerly direction its clocks are put back by one hour for every 15° of longitude, so that it will return to its home port with its calendar one day behind the date ashore.

To avoid these discrepancies the International Date Line was introduced, roughly following the 180° meridian, but deviating slightly from it to avoid passing through land. Ships and aircraft crossing the line in an easterly direction gain a day, so that if the date is April 10 it is changed to April 9, whilst those crossing it in a westerly direction lose a day, in which case April 10 is changed to April 11.

Great Circles

The shortest distance between any two points on a globe can be found by holding a piece of thread between them and drawing it taut against the surface of the globe. If the thread could be continued in the same direction round the globe it would form a great circle, which means that it is one of the largest circles that it is possible to draw on the globe.

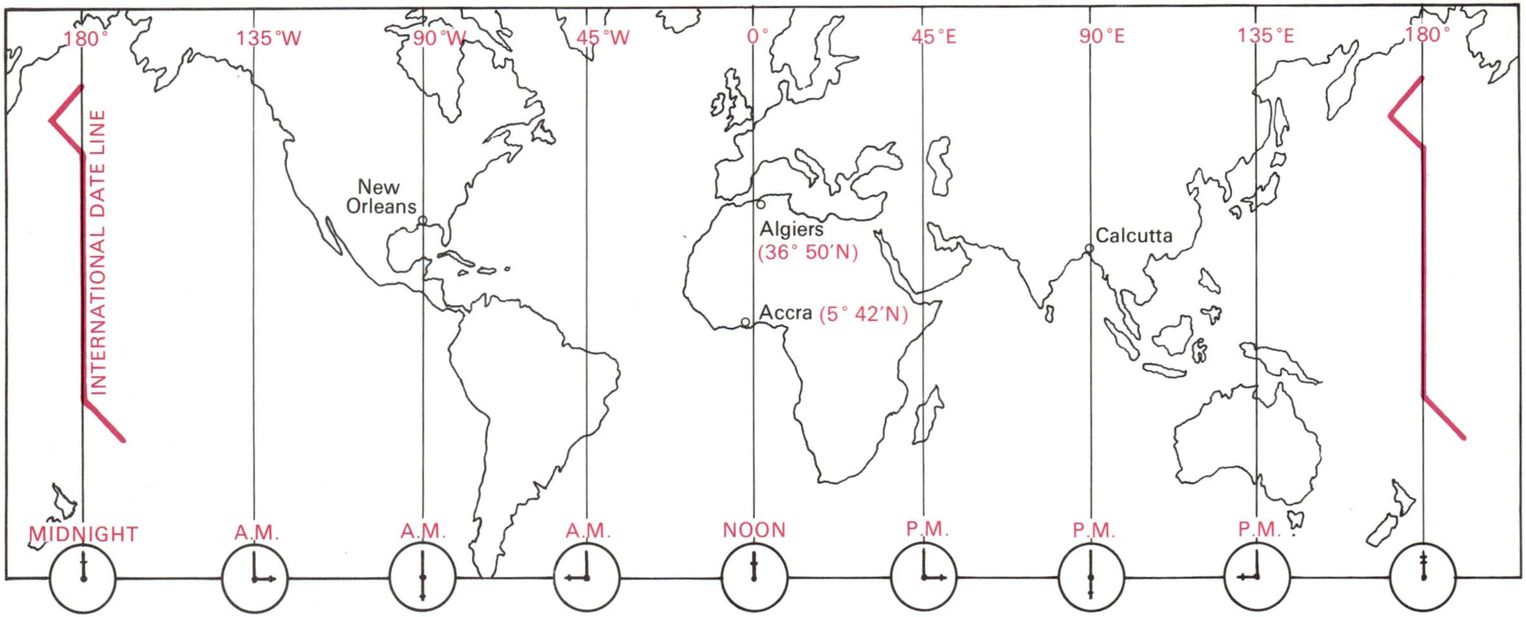

Figure 3.6 Longitude and time
(*a*) What time would it be at Calcutta and New Orleans if it were 6.30 p m at London? (*b*) If the navigator of a ship finds with a sextant that the Sun has reached its highest point in the sky (i.e. that it is local noon), and notes that the chronometer shows 9.30 a m GMT, what is the ship's longitude? (*c*) What is the distance from Accra to Algiers? Explain your calculation.

The shortest distance between any two points on the Earth's surface is therefore along part of the great circle which passes through them. These great circles are used by ships and aircraft wherever possible. For example shipping routes between the United Kingdom and Canada curve northwards in the centre of the North Atlantic Ocean, though they are often prevented from following exactly a great circle route by the presence of icebergs round the south coast of Greenland.

LUNAR AND SOLAR ECLIPSES

A lunar eclipse occurs when the Earth passes between the Sun and the Moon, casting its shadow on the Moon. This can only happen at the time of Full Moon, but does not happen at every Full Moon because the plane of the Moon's orbit is inclined at about 5° to the plane of the Earth's orbit. If the Moon is completely obscured there is said to be a total eclipse, and if partly obscured, a partial eclipse.

A solar eclipse occurs when the Moon passes

between the Earth and the Sun, and casts its shadow on the former. This can only occur at New Moon, but does not occur at every New Moon, again owing to the inclination of the planes of the Earth's and Moon's orbits. There may be a total or partial eclipse, but a total eclipse is visible only along a very narrow belt. Many countries send expeditions to areas where total eclipses are to be visible, mainly to study the *corona*, the fringe of radiant light produced by the gaseous atmosphere of the Sun, which is visible only then. The next total eclipse to be visible from Great Britain will be in 1999.

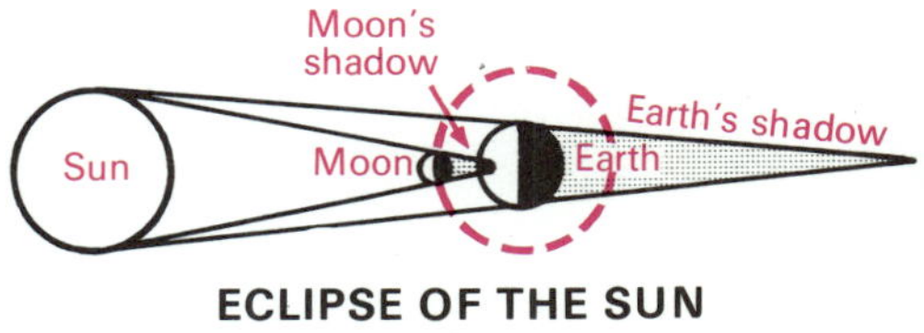

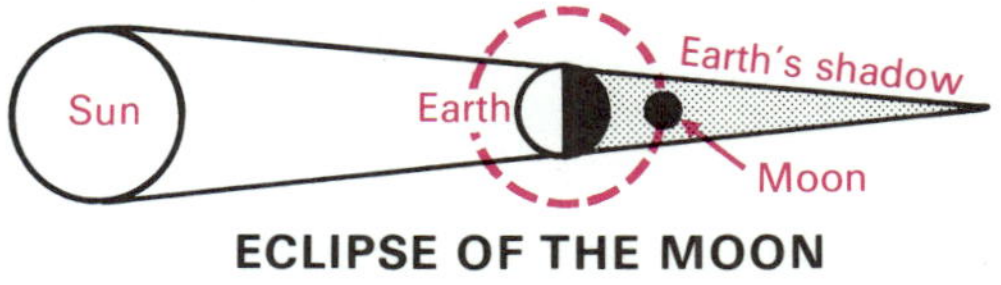

Figure 3.7 Eclipses of the Sun and Moon

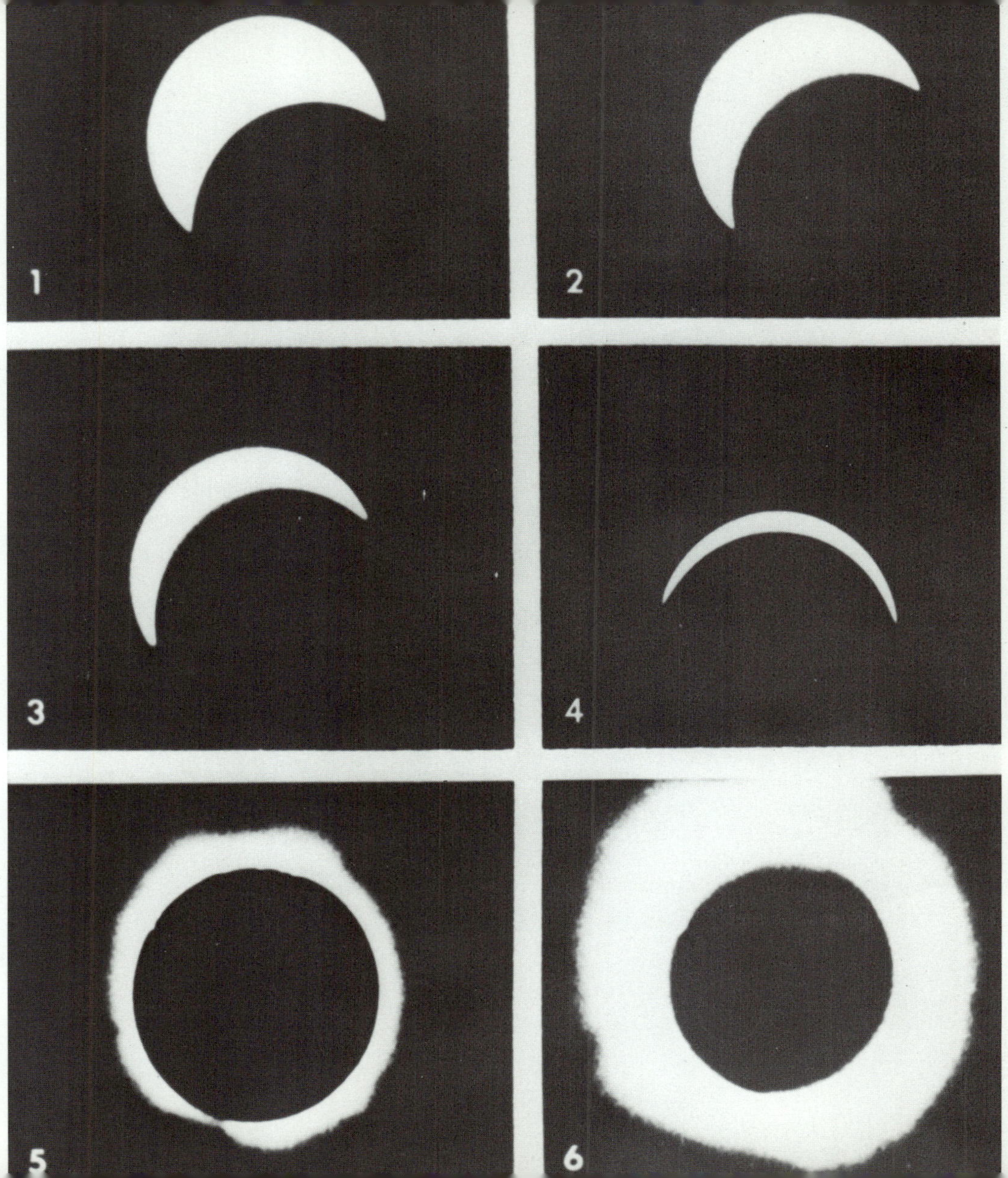

Solar eclipse (*United Press International*)
A sequence of photographs taken from a jet airliner over the east coast of the United States on 7 March 1970.

◀ **Earthrise** (*U S Information Service*)
The Earth, as seen from the Moon on the Apollo 8 mission.

Exercises

1 On an outline map of the world:

(*a*) Draw and name the following lines of latitude: the Equator, the Tropics of Cancer and Capricorn, the Arctic and Antarctic Circles. Against each line write its latitude.

(*b*) Draw and name the Prime Meridian and the International Date Line.

(*c*) Print the following along the correct lines of latitude:
Sun overhead on June 21
Sun overhead on December 21
Sun overhead on March 21 and September 22.

(*d*) Shade lightly the areas between the Arctic Circle and the top edge of the map, and between the Antarctic Circle and the bottom edge. Print the following over the correct areas:
Continual daylight on December 21
Continual darkness on June 21
Continual daylight on June 21
Continual darkness on December 21

2 The Phases of the Moon

In a darkened room stand about 5 metres away from a weak light (a candle is sufficient) and hold a black ball at arm's length in front of your face. Imagine that the light represents the Sun, the ball the Moon and your head the Earth. Now turn slowly round to the left, noting how much of the lit up part of the ball is visible to you.

Which phases of the Moon are represented (*a*) when you are facing the light, (*b*) when you are looking sideways to the light and (*c*) when you are looking away from the light? If the ball obscures the light in (*a*), or if your shadow falls on the ball in (*c*), what do these situations represent?

3 A Model to Demonstrate the Seasons

Use a table lamp (without the shade) to represent the Sun and four small portable globes to show the Earth in different positions round the Sun, with their axes pointing in the same direction as in figure 3.4. The light should be switched on in a darkened room and the direction of the Sun's rays demonstrated with a ruler. By rotating the globes the lengths of day and night in different parts of the World can be estimated.

On which dates is the Sun's altitude greatest and least, where you live? Using a piece of cotton, work out the approximate lengths of daylight and darkness on June 21, September 22, December 21 and March 21, for your home area.

Chapter 4

The Atmosphere

The Earth is surrounded by a thin layer of gases called the atmosphere, which is attracted towards it by the force of gravity. These gases include nitrogen (78 %), oxygen (21 %), argon, carbon dioxide and water vapour, but minute droplets of water and particles of dust are also present. The dust is derived from a variety of sources, such as smoke, volcanoes, meteorites, salt from sea spray and many types of micro-organism.

The atmosphere can be divided into three zones:

1 Troposphere. This zone is the most familiar to us. It extends upwards for about 11 kilometres from the Earth's surface, somewhat higher round the Equator and lower near the Poles. Both the air temperature and pressure tend to fall with increasing height but, weather conditions being often unstable; there are local fluctuations in temperature and pressure, accompanied by variable winds, cloud and rain.

2 Stratosphere. Above the Troposphere lies the Stratosphere, extending upwards to a height of about 80 kilometres. Here the temperature remains fairly constant, there is little water vapour and the air becomes increasingly rarefied. Jet planes fly in the Stratosphere since they are more efficient at low atmospheric pressures.

3 Ionosphere. Higher still is the Ionosphere, a zone of very thin air in which are found the electrically charged layers that reflect wireless waves back to the Earth. Another feature of the Ionosphere is the *aurora*, a brilliant display of lights often seen in high latitudes, and which is the result of currents of electricity produced by streams of electrified particles from the Sun.

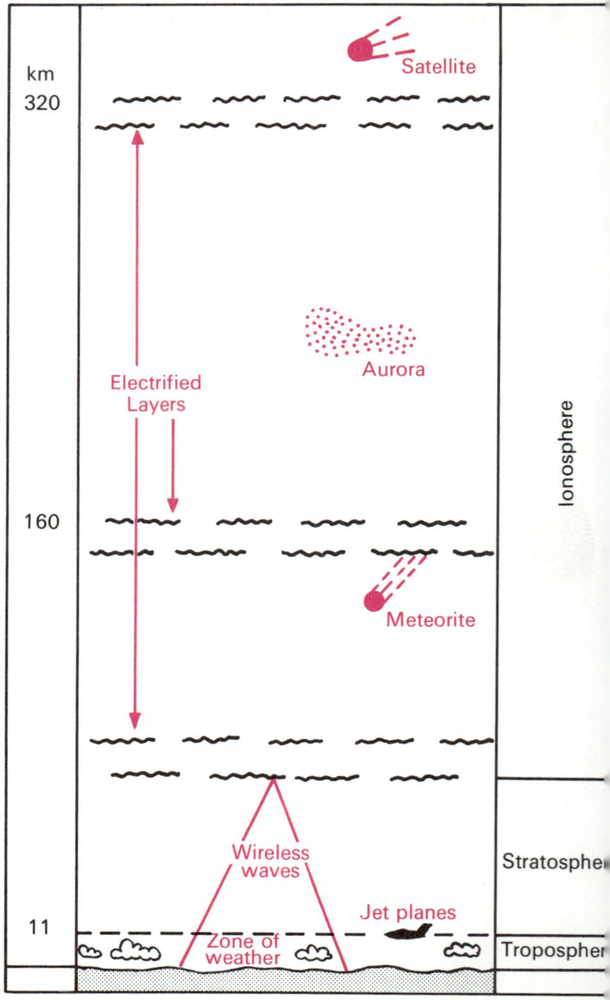

Figure 4.1 Section of part of the atmosphere

In the past most of our knowledge of the atmosphere has been limited to the Troposphere, but with the development of instrumented rockets and Earth satellites much progress has recently been made in upper atmosphere and space exploration. It is now known that the atmosphere extends upwards, in an extremely rarefied form, for over 1000 kilometres. Variations in the speed of satellites have shown that the atmosphere expands and contracts, rather like a gigantic balloon, and that such movements are closely related to eruptions on the Sun. These sunspots, as they are often called, cause magnetic storms (disturbances of the Earth's magnetic field) and interfere with radio communications.

TEMPERATURE

Our main source of heat is the Sun. From its glowing mass radiant energy is thrown far out into space. The Earth, at a distance of 150 million kilometres, is able to intercept only a tiny fraction of this energy. Some is absorbed or reflected by the atmosphere, and the rest reaches the Earth's surface in the form of heat and light. At night-time, on the other hand, the Earth itself loses heat by radiation into space.

The degree of hotness, or temperature, can be measured with a thermometer. Prior to 1962 the Fahrenheit scale was in general use for air temperature, but this was replaced by the Celsius or Centigrade scale. The actual temperature at any place depends principally on the following factors:

1 Angle of the Sun's Rays. Because of the curvature of the Earth the Sun's rays do not fall at the same angle everywhere. In figure 4.2 equal amounts of heat from the Sun are shown as three beams. It will be seen that the beams falling near the North and South Poles have to heat a larger surface area and travel farther through the atmosphere, and therefore have less heating effect than the one that falls over the Equator. Similarly the difference in the heat received in summer and winter at any one place is largely due to the difference in the angle of inclination of the Sun's rays.

Figure 4.2 The heating of the earth

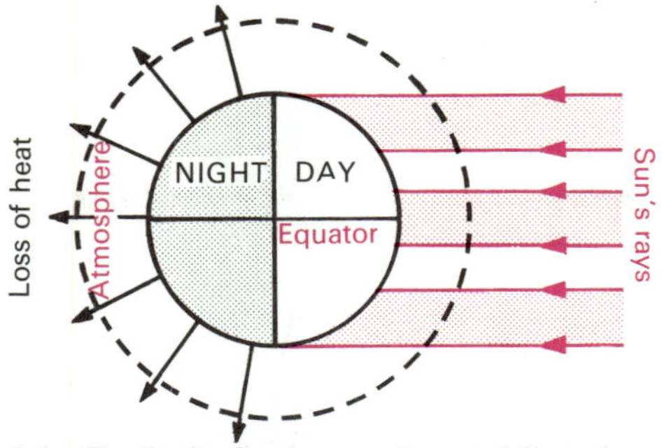

2 Length of Daylight. Variations in the length of daylight also contribute to the difference between the summer and winter temperatures. In summer the long days and short nights produce a general rise in temperature, whilst in winter the long nights and short days cause an overall loss of heat and consequent fall in temperature.

3 Effect of the Sea. Under similar conditions water takes longer than land to heat up and to cool down, so that in general the sea is cooler than the land in summer and warmer in winter. The sea thus affects coastal areas and islands such as the British Isles, causing them to have an equable climate, neither very warm in summer nor very cold in winter. On the other hand places a long way from the sea, such as Central Siberia, have an extreme climate, with warmer summers but much colder winters. The temperature of the sea depends partly on ocean currents; for example, the seas round the British Isles are warmed by the North Atlantic Drift, which originates in the Gulf of Mexico. In contrast, the seas off Eastern Canada are cooled by the Labrador Current, which originates in the Arctic Ocean.

4 Influence of Winds. Winds may also have a warming or cooling effect. Thus the British Isles are mostly under the influence of south-westerly winds blowing from the Atlantic Ocean, which keep our islands cool in summer and mild in winter. Sometimes in winter, however, east or north-east winds may blow from the Continent, causing temperatures to fall well below freezing point.

5 Water and Dust in the Atmosphere. When these are present, they tend to blanket the Earth and lessen variations in temperature between day and night by absorbing some of the Sun's heat by day and reducing the loss of heat from the Earth by night. Thus the frequent cloud cover over the British Isles is one reason for the absence of extremes of temperature there.

6 Altitude. The height of any place above sea level has a considerable effect on the temperature, which falls at an average rate of about *1 degree Celsius per 160 metres ascent*. This is because the atmosphere receives most of its heat from the Earth, mainly by the processes of conduction and convection, rather than directly from the Sun.

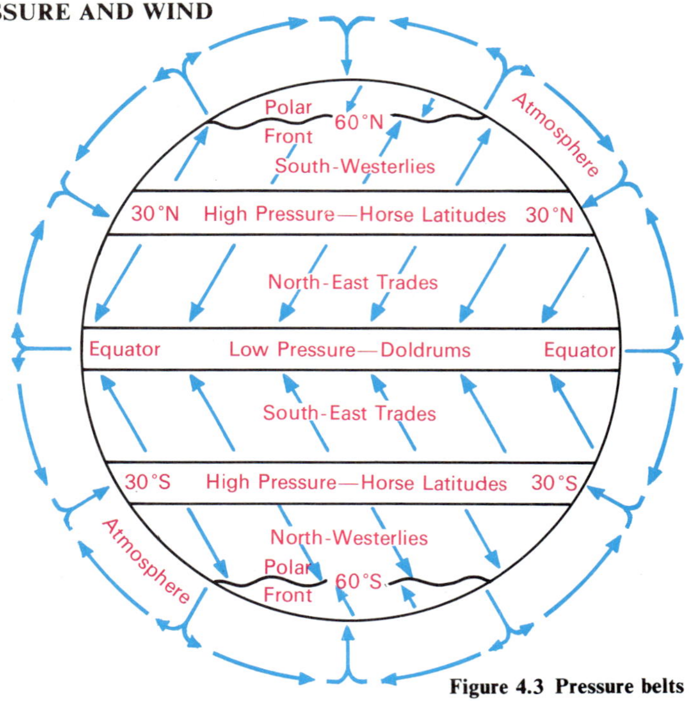

Figure 4.3 Pressure belts and prevailing winds

The pressure of the atmosphere at sea level is sufficient to support a column of mercury about 76 cm high in a barometer. But pressure is usually measured in millibars rather than in centimetres, 1,012 millibars being equivalent to 76 cm of mercury.

There are considerable variations in pressure in different parts of the world. Owing to high temperatures along the Equator the air near the ground is heated. It expands and rises, producing a low-pressure belt called the *Doldrums*. The rising air cools and spreads out, until it sinks again at about latitudes 30° north and south, to form the high-pressure belts known as the *Horse Latitudes*. The Polar regions, owing to the accumulation of cold, heavy air, also tend to be areas of high pressure.

These high- and low-pressure belts are mostly areas of comparative calm, but between them blow the prevailing winds. Wind is a movement of air from areas of high pressure to areas of low pressure. It is also affected by the Earth's rotation, being deflected (presuming that you have your back to the wind) to the right in the northern hemisphere and to the left in the southern. Thus in the northern hemisphere winds blow from the Horse Latitudes to the Doldrums (that is, from high pressure to low pressure), but are deflected to the right to form the *North-East Trades*; other winds blow from the Horse Latitudes towards the Poles, but are similarly deflected to form the *South-Westerlies*. In the southern hemisphere, where the deflection is to the left, the *South-East Trades* and *North-Westerlies* are produced. The latter blow strongly over great expanses of ocean, in contrast to the South-Westerlies which are more variable owing to the presence of great land masses.

From the Polar regions cold north-east and south-east winds blow outwards and come up against the Westerlies in a highly disturbed zone, between 50° and 70° north and south. The surface of contact between these warm and cold winds is called the *Polar Front* and along it the warm air tends to rise over the cold air, producing an irregular low-pressure belt.

This description of the world pressure and wind belts give the impression that they are more or less stationary. In fact there is a pronounced seasonal movement, following the overhead Sun, northwards in the northern summer and southwards in the southern summer. This carries the pressure belts to a position about 8° farther north in June and 8° farther south in December than is shown in figure 4.3. A further complication is produced by the unequal distribution of land and sea, which modifies both the pressure and winds.

Some reference must be made here to local

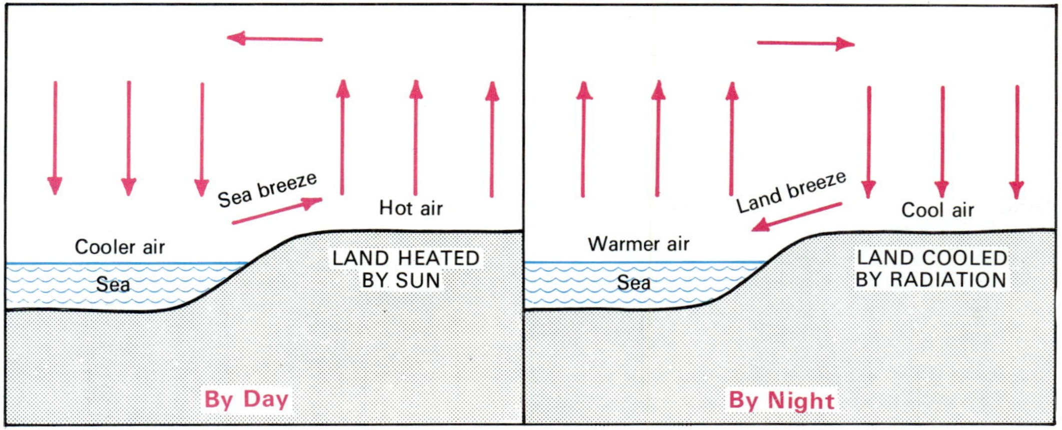

Figure 4.4 Land and sea breezes

winds. In coastal regions, during calm, clear weather, *land and sea breezes* develop. By day the land heats up more rapidly than the sea and the air over the land expands and rises, causing a low-pressure area towards which the cooler air moves from over the sea. By night the land cools more rapidly than the sea and cold, heavy air from over the land tends to move out over the sea, whilst the warmer air there tends to rise.

Mountain and valley breezes develop under similar conditions. The cold mountain air descends into the valleys and adjoining plains during the night, but in the daytime, since the mountains are the first to be heated by the Sun, the warm air there rises, causing a wind to blow from the valleys towards the mountains.

WATER IN THE AIR

The greatest part of the Earth's surface is covered by water. Even on the land, lakes, rivers and swamps cover a considerable area, whilst much of the soil and vegetation contains water. From these surfaces water is continually entering the air in the form of water vapour, the process being known as *evaporation*. But at a given temperature there is a limit to the amount of water vapour which the air can hold. When this is reached the air is said to be *saturated* and water will cease to evaporate unless (*a*) a wind is blowing, bringing in fresh unsaturated air, or (*b*) the temperature rises, for the higher the temperature, the more water vapour the air can hold.

The weight of water vapour in a given volume of moist air divided by the weight required to saturate this volume at the same temperature is called the *relative humidity*, and is generally expressed as a percentage, saturation point being 100 %. When moist air is cooled, the temperature at which it becomes saturated is known as the *dew point*; further cooling will cause some of the water vapour to change back to water. This last process is called *condensation*, and may result in the formation of cloud, mist (or fog), dew (or hoar frost), rain or snow.

Cloud

The cooling of the air generally causes some of the water vapour in it to condense into a cloud of tiny droplets of water (or ice), which remain suspended in the air. Dust in the atmosphere plays an important part in this process, each droplet of water tending to condense round a nucleus of solid matter. Some of the commonest types of cloud are:

1 Cirrus. These fibrous, white, ice clouds form at heights above 6000 metres. They may resemble 'mares' tails' and are often a sign of the approach of bad weather. The photograph on page 44 shows cirrus.

2 Stratus. Much lower, generally below 2000 metres, this forms a uniform grey sheet, which may bring light rain or drizzle. A denser, darker form of it, called nimbostratus, produces heavy rain or snow.

3 Cumulus. These have a low, level base, but can tower to great heights, with white or light grey cauliflower-shaped heads. They are caused by ascending air currents, and may produce showers of rain. On warm, sunny days, shallow fair weather cumulus may develop. The very dark cumulonimbus, sometimes with anvil-shaped heads, gives heavy, thundery showers.

In addition to these main types there are a number of others, details of which are given in the International Cloud Atlas.

Rain

The tiny droplets of water that make up clouds may unite to form larger drops, which fall as rain. In a similar way ice crystals may grow to form snowflakes. There are three principal causes of rain:

1 Relief Rain. This is caused by high ground lying in the path of a wind containing water vapour. The ascending air expands and cools, producing cloud, followed by rain or snow, especially on the windward slopes and round mountain summits. When the air descends on the leeward side it becomes compressed and warmed, so that rain does not fall. This area is termed the *rain shadow*. Relief rain is common in mountainous regions, for example the rainiest place in England is Borrowdale, in the Lake District, where about 4200 mm per annum is received.

2 Convectional Rain. This is caused by the heating of the ground by the Sun. The air near the ground is warmed and this causes it to expand and rise as a convection current. It cools as it rises and clouds form, developing into cumulonimbus and producing heavy rain. Sometimes raindrops are carried by a powerful updraught to levels where the air temperature is well below freezing point, and they become hail. The hail may commence to fall, only to be lifted again and acquire a new layer of ice. Some large hailstones, when cut open, are found to be composed of several concentric layers. Thunder and lightning, which may accompany these downpours, are due to the escape of static electricity from the clouds, where it has accumulated. This electricity is derived partly from the break-up of large raindrops in the lower part of the cloud, and partly from friction between ice crystals. Convectional rain is one of the commonest types of rain in tropical areas, and occurs occasionally in the British Isles, particularly in summer.

3 Cyclonic Rain. This type of rain is produced by the temporary low-pressure systems known as *depressions* or *cyclones*. Tropical cyclones are usually more violent than those in temperate latitudes, and may take the form of *typhoons* or *hurricanes*, with strong winds blowing spirally inwards. In temperate regions, such as the British Isles, they are called depressions, and develop along the Polar Front, where warm, moist air is forced to rise over cold air. The heaviest rain occurs along the fronts. Depressions are fully described in Chapter 5.

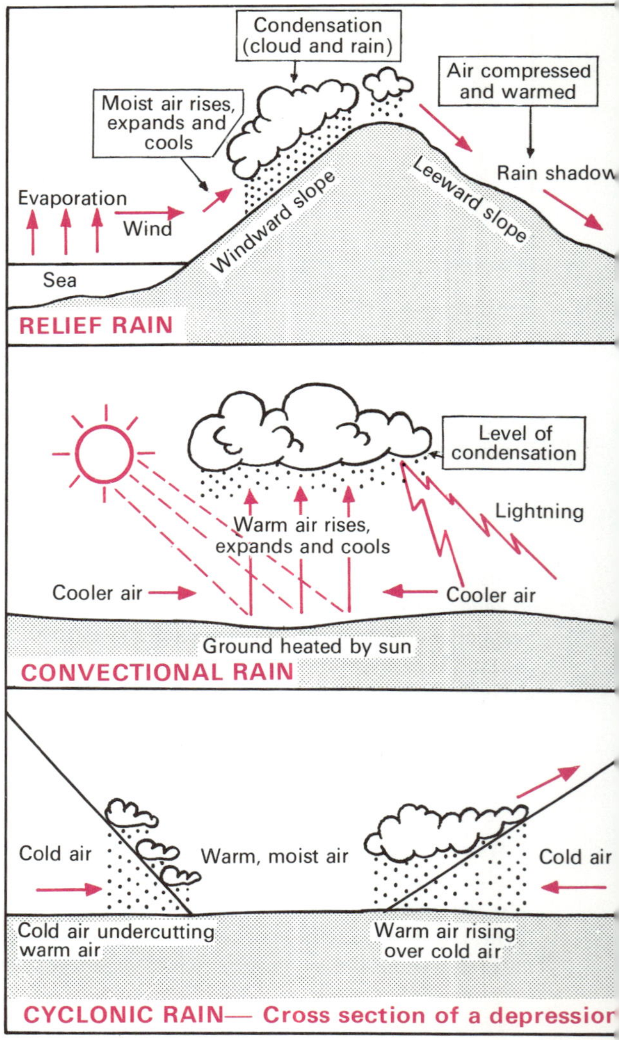

Figure 4.5 The causes of rain

left **Radiation fog over Hitchin;** *right* **Storm over Philadelphia** (*Aerofilms*)

Describe the weather in these two photographs, and state the conditions needed to produce these types of weather.

Mist and Fog

These are a form of cloud near the ground. When the visibility is under 1 kilometre the condition is called fog, and when it is between 1 and 2 kilometres we speak of a mist. Mist and fog will form, when, for example, a warm autumn day is followed by a clear, calm night. The warm day produces an abundance of water vapour, and during the night the earth loses heat by radiation, cooling the air near the ground until dew point is reached, at which stage mist begins to form, later developing into fog. The fog may be particularly thick in hollows and valleys because of the tendency of cold air, which is denser than warmer air, to pour down from the surrounding hills. The cold air near the ground may have warmer air above it; this reversal of the

Cirrus and cirro-stratus (*F R Dobson*)

Cumulus (*F R Dobson*)

Figure 4.6
Cause and effect of an inversion of temperature

normal situation, in which temperature falls with height, is known as an *inversion of temperature*. The warm air acts like a lid, since cold air cannot ascend through warmer air. Over cities and towns smoke is unable to escape and combines with the fog to form an unhealthy mixture, sometimes called *smog*. In summer, early morning mist usually disappears soon after sunrise, but in winter, when the Sun's rays are weak and there are few hours of daylight, fog may persist for days, until dispersed by wind.

Mist and fog may be formed in quite a different way, by the mixing of warm and cold air. An example of this is found off the east coast of Newfoundland, where two ocean currents, the warm Gulf Stream and the cold Labrador Current, come together. The warm, moist air from over the Gulf Stream has its temperature lowered below dew point by mixing with the cold air, so that dense fogs are formed. The early summer fogs along coastal areas in Britain are another example, being caused by warm sea winds blowing over belts of cold water off the coast.

Dew and Hoar Frost

The chilling of the earth at night can also induce condensation in the form of dew on blades of grass or on the soil. When dew point is below freezing point hoar frost is formed. This is white in colour because air is trapped between the particles of ice owing to the rapid freezing. Glazed frost, more

Cumulonimbus with rainbow (*F R Dobson*)
Rainbows are produced by the refraction and internal reflection of sunlight in minute droplets of water in the air.

Rime and hoar frost (*F R Dobson*)

difficult to see and therefore more dangerous on the roads, is produced in a different way by the slow freezing of water from which air is excluded. The rime which forms on the windscreens of cars is caused by super-cooled water droplets (which have remained in a liquid state at a temperature well below freezing point) changing to ice when they come into contact with the car. In a similar way rime is formed by wind blowing the droplets against trees or fences.

The Hydrological Cycle

Water is continually circulating between sea, air and land. The sequence of events is called the water cycle or the *hydrological cycle*. Moisture is *evaporated* from the surface of the ocean by air passing over it. When the air containing water vapour reaches the land, it rises and, as explained on page 42, condensation occurs and clouds form. Eventually, it rains.

Some of the rain falls straight into rivers and flows directly to the sea. Some rain lies on the land or is trapped by vegetation and is eventually evaporated into the air again. Much of it soaks into the soil. Some of this is taken up by plants and is passed back into the air by transpiration. The rest passes through the soil to the rivers or down into porous rock and more slowly to the rivers. The water finally reaches the ocean, and the whole cycle starts again.

Figure 4.7 shows the general features of the hydrological cycle. In any particular place the importance of each part may vary according to how much rainfall occurs, how much vegetation there is, or how much building there is on the land.

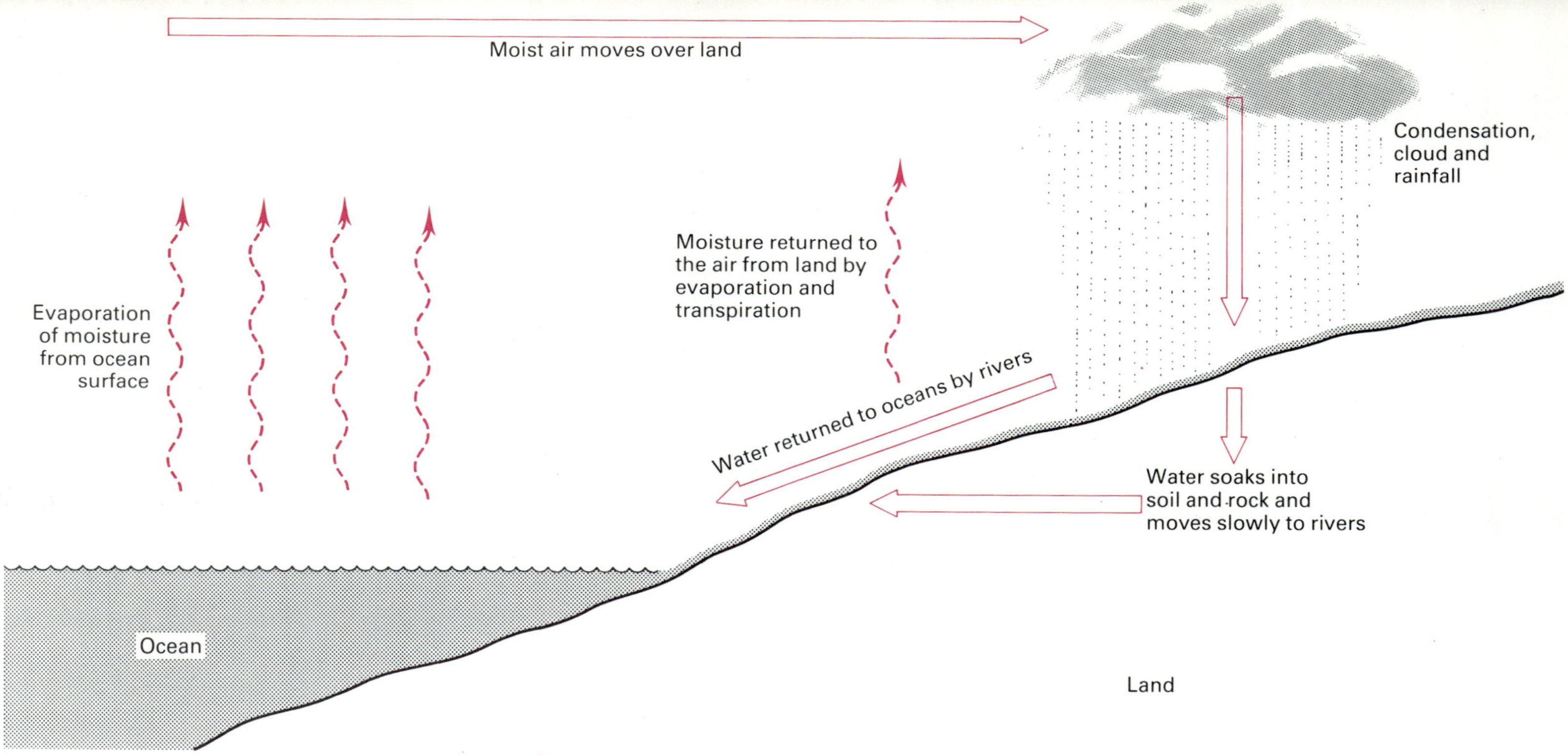

Figure 4.7 The hydrological cycle

Exercises

1 (*a*) Why is it possible to enjoy the warmth of the Sun, surrounded by snow and ice, in high mountains such as the Alps?

(*b*) Both heat waves and long periods of severe frosts are rare in the British Isles. Explain fully the reasons for this.

2 (*a*) Why is the inside of your bedroom window often covered with moisture on a cold morning?

(*b*) Why do wet clothes dry more quickly (i) in summer than in winter, and (ii) on a windy day than on a calm day?

(*c*) How could smog over our great cities be avoided?

3 Study figure 4.7 carefully.

(*a*) How would the hydrological cycle be affected in areas with a heavy winter snow cover?

(*b*) What effect would the creation of a huge reservoir have on the hydrological cycle in an area?

Chapter 5

The Measurement of Climate and Weather

Weather station at London Airport (*Crown Copyright*)

Measuring rainfall (*Crown Copyright*)

Figure 5.1 Lay-out of a weather station

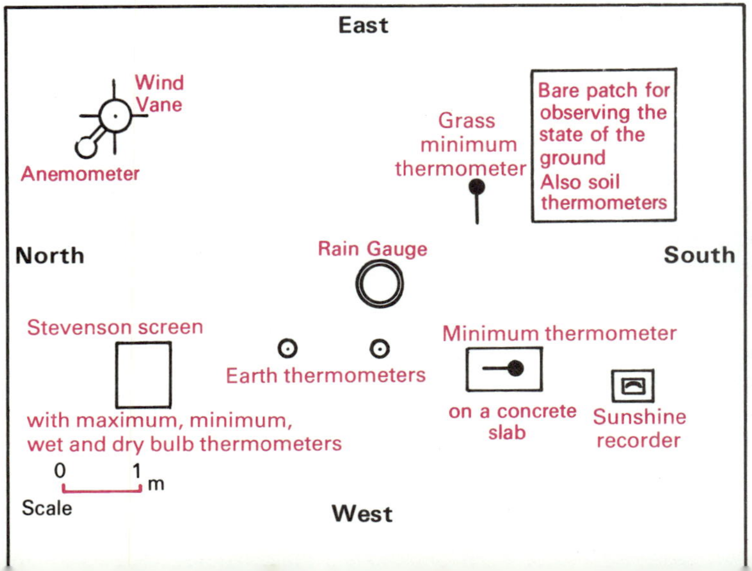

Reading the mercury barometer (*Crown Copyright*)

In past centuries our knowledge of climate and weather was based on rough, casual observation and experience. Today we make organised and accurate measurements, using specially designed instruments. There are large numbers of fully equipped weather stations in the United Kingdom. Some of these are called *meteorological stations* and others *climatological stations*, but they have the same basic equipment, which must conform to the standards laid down by the British Meteorological Office.

A weather station should occupy a level or nearly level piece of ground and the instruments should be kept in a fenced enclosure measuring 10 metres by 7 metres. The site should be well away from buildings and trees. Detailed instructions for setting up a station can be obtained from the Meteorological Office. The following are the most important instruments:

Maximum and Minimum Thermometers. These are kept in a *Stevenson screen*, which is a wooden box, standing about $1\frac{1}{4}$ metres from the ground, painted white and having double louvered sides. The maximum thermometer registers the highest temperature during the day. It contains mercury and depends for its action on a small constriction in the bore of the tube. As the temperature rises the mercury is forced through this constriction but does not return when the temperature falls. The thermometer is reset by shaking it downwards, with a steady swinging motion of the hand. The minimum thermometer registers the lowest temperature during the day and contains alcohol. Its movable index is pulled back as the temperature falls. It is reset by tilting the bulb upwards.

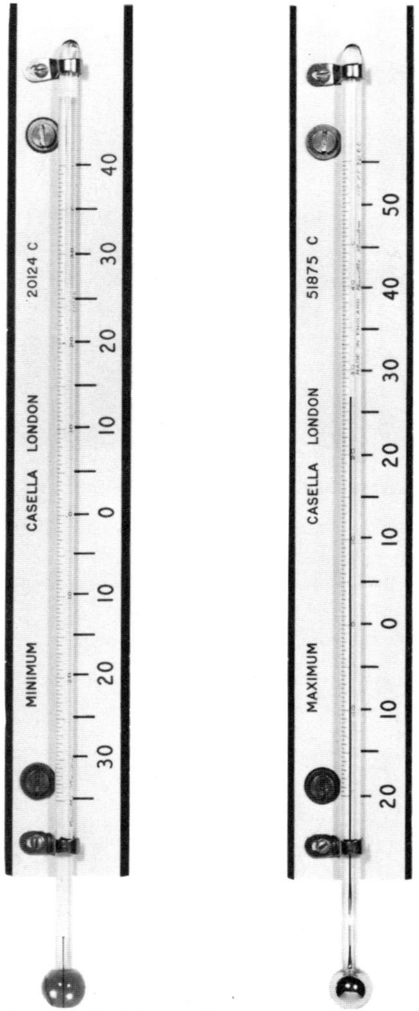

Maximum and minimum thermometers (*C F Casella*)

There is also a combined type of thermometer, known as Six's Maximum and Minimum Thermometer. This is U-shaped and contains both mercury and alcohol, with small indices to indicate the highest and lowest temperatures.

Wet and Dry Bulb Thermometers. These are also kept in the Stevenson screen. The dry bulb thermometer contains mercury and measures the air temperature. The wet bulb thermometer is of a similar type, but its bulb is covered with muslin which is kept moist by means of a wick, the end of which is immersed in a jar of distilled water. The temperatures shown on the two thermometers differ slightly, due to the evaporation of water from the wet bulb. This difference in temperature, or depression of the wet bulb reading, is used to find the relative humidity from a table such as the one at the end of this chapter.

Grass Minimum Thermometer. This is similar to the minimum thermometer already described and is used to measure the lowest temperature reached during the night over grass freely exposed to the sky. It is placed on two Y-shaped pieces of wood, between 2 and 3 cm from ground level, just touching the blades of grass.

Barometer. The most accurate instrument for measuring the atmospheric pressure is the *mercury barometer*. It consists of a glass tube containing mercury, with its lower end held in a steel cistern also containing mercury, the upper end of the tube being sealed. Air pressure on the mercury in the cistern supports the weight of the mercury in the tube, the height of which indicates the pressure. Another type of barometer, the *aneroid barometer*, contains a small corrugated metal cylinder with a

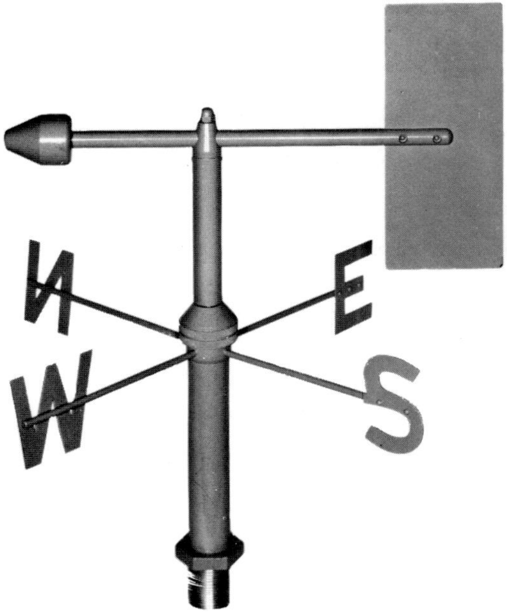

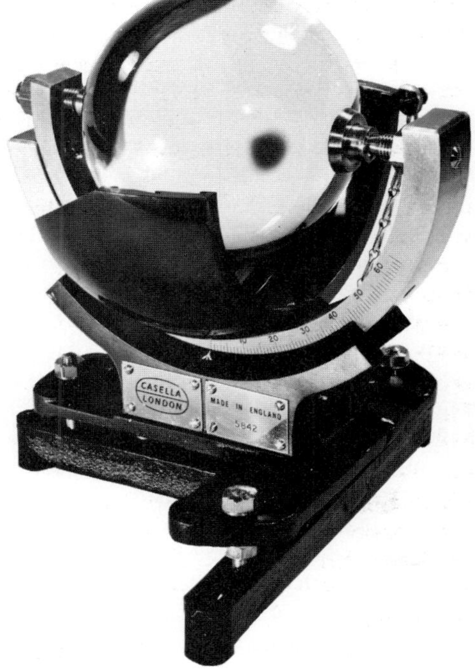

left **Wind vane;** *centre* **Anemometer;** *right* **Sunshine recorder** (*C F Casella*)

partial vacuum inside. Variations in air pressure cause the lid to move, the movement being recorded on a dial by a moving needle. If a continuous record of the pressure is required a *barograph* is used. This incorporates an aneroid barometer and a clock drum on which a weekly chart is fitted. A pen operated by the barometer records the pressure on the chart as the drum revolves. Air pressure is virtually the same inside a building as outside, so that most barometers are kept indoors, but they should, if possible, be out of reach of direct sunlight.

Rain Gauge. This is placed in the centre of the enclosure and consists of a funnel, fixed over a collecting jar, in an outer vessel. The upper part of the funnel should be 30 centimetres above grass level. A measuring cylinder completes the equipment. The rain gauge measures all forms of precipitation including rain, hail, snow, sleet and dew. If the funnel is blocked by snow or hail these can be melted by applying gentle heat.

Wind Vane and Anemometer. The wind vane should be mounted on a mast at least 3 metres higher than any neighbouring buildings or trees. A movable arm, fixed above the compass points, indicates the direction of the wind. The anemometer should be 6 metres higher than neighbouring buildings or trees. It measures the wind speed by means of three metal cups which are rotated by the wind. There are various methods of indicating the wind speed; in one of these the anemometer is fitted with electrical contacts connected in series with a battery and buzzer. By counting the number of buzzes over a given period of time (according to the manufacturer's instructions) the wind speed can be found.

In the absence of an anemometer it is possible to estimate the wind speed, using the *Beaufort Scale*, as shown here.

Sunshine Recorder. This should be set up on a rigid support where there is nothing to obstruct the Sun's rays at any time of the day or year. It records the number of hours of sunshine each day. One type has a spherical lens which causes the Sun's rays to scorch a line on a card; another simpler type has a metal cylinder with two slots through which the Sun's rays pass to leave a trace on sensitised paper.

Log Book. The instruments at a fully staffed weather station are read several times a day, but at a school weather station it is sufficient to take the readings once a day, preferably at 9 a m. These readings are then entered in a log book, part of a page of which is shown in figure 5.2. A number of weather observations are also made. These include the type and amount of cloud (in eighths), the visibility in kilometres, and any special weather phenomena, such as thunderstorms, dew, hoar frost or the depth of snow.

It is also useful to record weather readings graphically, with continuous lines to show the temperature, humidity and pressure, and a series of vertical columns for rainfall. Wind directions may be shown on a wind rose.

Beaufort no.	Wind	Speed in knots	Commonly observed effects
0	Calm	0	Calm, smoke rises vertically.
1	Light air	1–3	Direction of wind shown by smoke drift, but not by wind vanes.
2	Light breeze	4–6	Wind felt on face; leaves rustle; ordinary vane moved by wind.
3	Gentle breeze	7–10	Leaves and small twigs in constant motion; wind extends light flag.
4	Moderate breeze	11–16	Raises dust and loose paper; small branches are moved.
5	Fresh breeze	17–21	Small trees in leaf begin to sway; crested wavelets form on inland waters.
6	Strong breeze	22–27	Large branches in motion; whistling heard in telegraph wires; umbrellas used with difficulty.
7	Moderate gale	28–33	Whole trees in motion; inconvenience felt when walking against wind.
8	Fresh gale	34–40	Breaks twigs off trees; generally impedes progress.
9	Strong gale	41–47	Slight structural damage occurs (chimney pots and slates removed).
10	Whole gale	48–55	Seldom experienced inland; trees uprooted; considerable structural damage occurs.
11	Storm	56–63	Very rarely experienced; accompanied by widespread damage.
12	Hurricane	64–71	—

Date	Barometer (millibars)	Thermometers °C					Humidity %	Wind	Visibility in kms	Cloud		Rainfall cms	Sunshine hours	Remarks
Dec.		Dry	Wet	Maximum	Minimum	Grass Minimum		Direction and Force		Amount	Type			
5	992	4·4	3·2	7·2	3·2	1·0	82	SW3	10	8/8	St	·5	Nil	—
6	996	5·4	3·8	8·8	0	-2·8	74	NW1	7	3/8	Cu	·1	Nil	Dew
7	1004	2·2	1·6	6·6	1·6	-3·8	90	NE2	1	0/8	—	Nil	4	Mist and hoar frost

Figure 5.2 Part of a page from a weather log book

HOW THE READINGS ARE USED

The Local Climate

The climate of any place is the average of the weather conditions that exist there. Climate can vary considerably, even within a small area, and a knowledge of the local climate is essential for the farmer, the architect, the water engineer and many industrial enterprises. It can be worked out quite simply from the daily readings taken at a weather station. The mean or average temperature each day is obtained by adding the maximum reading to the minimum and dividing by two. The mean monthly temperature for any month is obtained by adding the daily means and dividing the total by the number of days in the month. These monthly means should be plotted on a climate graph, so that the variations throughout the year can be seen at a glance. The seasonal range of temperature (the difference between the means of the warmest and coldest months) should also be found.

The monthly and annual rainfalls are obtained by simple additions and should also be shown graphically, by a series of columns similar to those in figure 5.3. Monthly and annual records may also be kept of the number of days with rain, the hours of sunshine, the number of observed thunderstorms, of days with fog and of the direction of the wind.

The figures for one year, although of considerable value and interest, do not give a reliable picture of the local climate, which must be based on readings taken over thirty or thirty-five years, in order to avoid the effects of any abnormal seasons.

Climate Maps

British weather stations send their climatic records to the Central Meteorological Office at Bracknell, in Berkshire, where, after careful analysis, they are plotted on maps. Temperature is shown by *isotherms*. These are lines joining places with the same mean temperature which in order to exclude the effect of altitude are adjusted to what they would be at sea level. They are usually drawn for January and July.

Rainfall maps employ a series of lines, called *isohyets*, joining places which have the same mean rainfall, the areas between the isohyets being usually coloured, as in figure 5.8. In the case of countries with unevenly distributed rainfall, such as India, there are normally two rainfall maps, each covering a six-month period, but for countries like the United Kingdom, where rain falls throughout the year, one annual rainfall map is often sufficient.

Atmospheric pressure is shown by means of *isobars*, which are lines joining places with the same sea-level pressure. The prevailing winds also are usually marked on isobar maps, which generally show the conditions in January and July.

Weather Maps and Weather Forecasting

Fully equipped meteorological stations are found at airfields, ports and other centres in Britain, Europe and North America, and a number of weather ships occupy fixed positions in the North Atlantic Ocean. These stations exchange weather information by teleprinter or radio, and plot the data on large charts which, when complete, are known as *weather maps*. Some of the symbols used on weather maps are shown in figure 5.4 and a

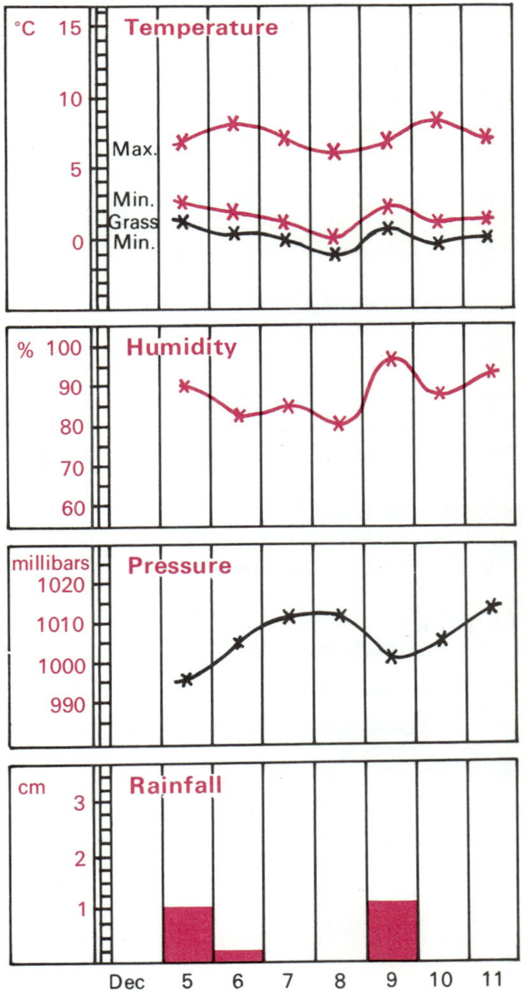

Figure 5.3 Weather graphs

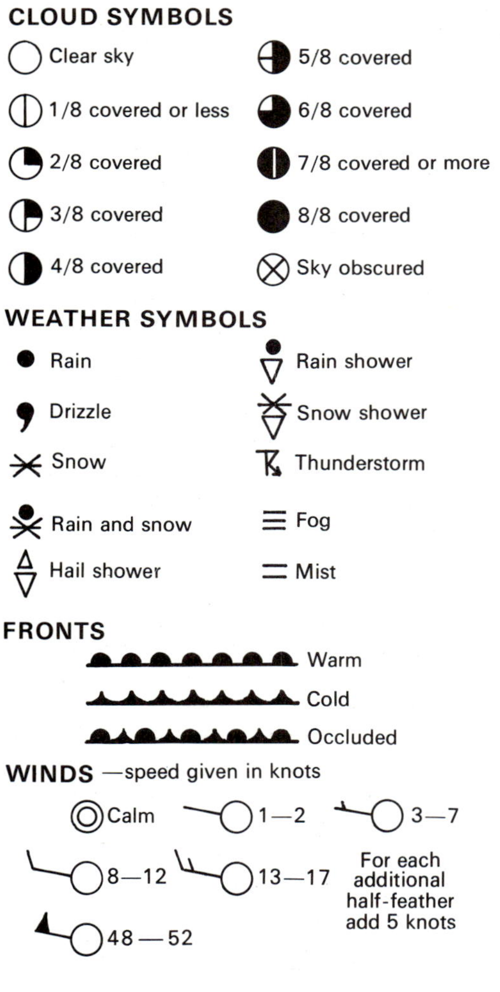

Figure 5.4 Symbols used on a weather map

simple example of a weather map is given in figure 5.7.

Weather forecasts are issued after a careful study of these maps. Forecasters are guided by two main factors: (*a*) their knowledge of the types of weather associated with the various pressure systems, and (*b*) their estimate of the likely direction and speed of movement of these systems. Forecasters must pay particular attention to the movement of the *fronts*, which are boundaries between air masses. The effect of fronts on the weather is explained in the next paragraph. The Central Meteorological Office at Bracknell issues general forecasts which are made available for shipping and the general public through the media of sound and television broadcasting and the press. In order to explain the principles of weather forecasting more fully some of the more important pressure systems which affect Britain's weather will now be considered.

Depressions

For most of the year Britain is under the influence of depressions, which usually reach our shores from the Atlantic and travel eastwards or north-eastwards across the country. They are areas of low pressure, with the winds blowing in an anticlockwise direction and slightly inwards towards the centre. A typical depression originates along the Polar Front, when warm and cold air masses become interlocked and a sector of warm air is almost surrounded by cold air. The leading edge of the warm sector is termed the *warm front* and along it the warm air moves slowly upwards over the cold air. The rear edge of the warm sector is the *cold front*, and here the cold air undercuts the warm air,

Figure 5.5 The birth of a depression

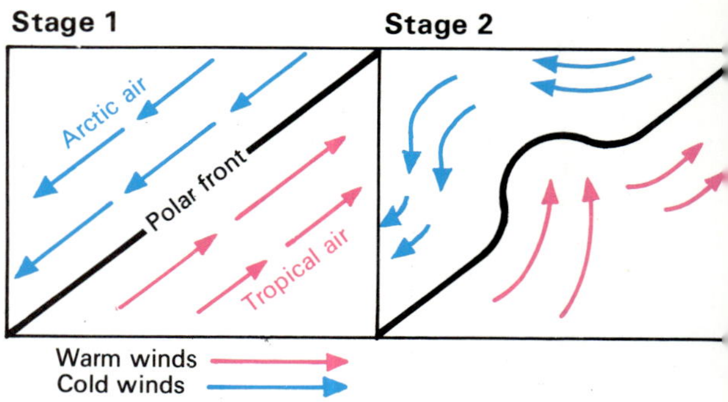

forcing it upwards. When the warm sector has been lifted completely off the ground the depression is said to be occluded. In this case the position of the lowest part of the warm air is marked on weather maps as an *occlusion*.

Figure 5.6 shows a depression, with a cross-section through it along the line XY. To an observer at Y the first sign of its approach will be some cirrus cloud. As the barometer begins to fall this will be replaced by a thin, whitish veil of cirrostratus, and the wind will increase in force from the south. With the approach of the warm front the clouds will darken to a low, grey nimbostratus, accompanied by drizzle, which will soon change to continuous heavy rain. After the warm front has passed the temperature will rise suddenly by a few degrees and there may be a lull in the rain, with lighter skies. The cold front will be accompanied by dense masses of cumulonimbus and heavy showers of thundery rain, with a distinct fall in temperature. The barometer will now begin to rise, with the wind

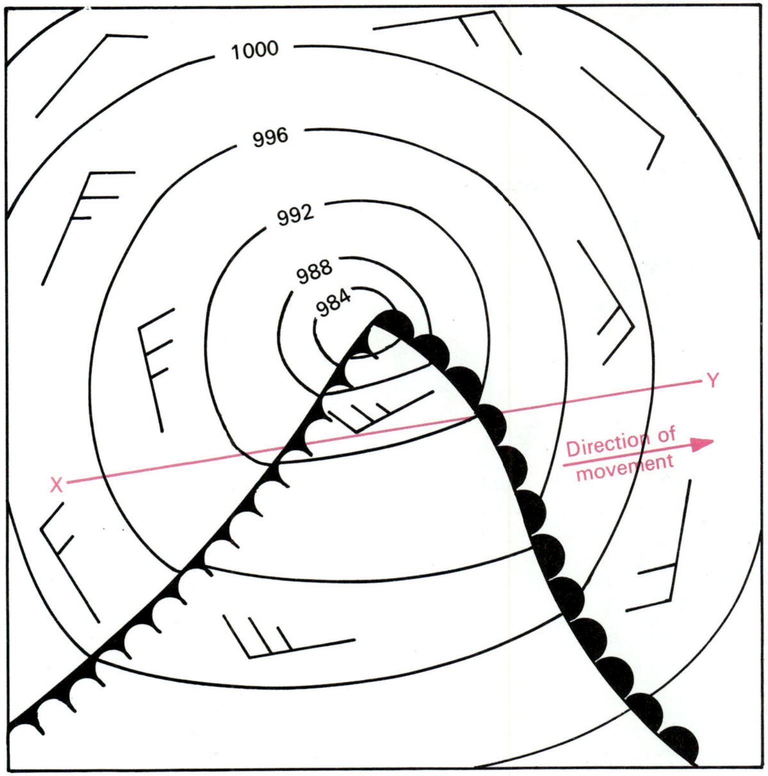

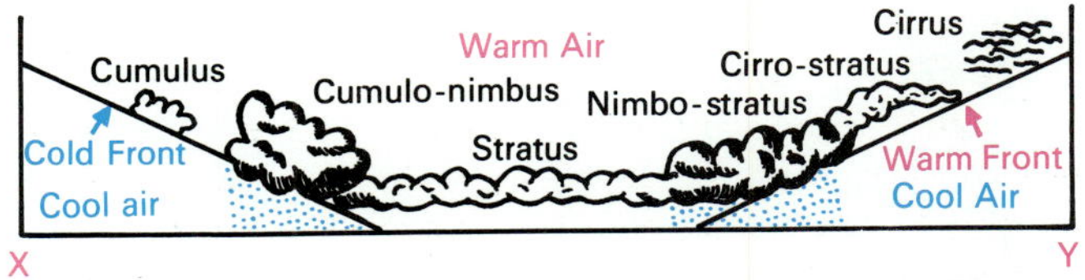

Figure 5.6 A frontal depression with a cross-section along the line XY

from the west or north-west. Cool, showery weather, with patches of blue sky, will follow, the showers tending to die out.

The occluded depressions are without fronts at ground level, although in many respects the weather they produce is similar to that accompanying a frontal depression. In some depressions the fronts may take the form of troughs, with the isobars bending abruptly and producing a sudden change in wind direction as the trough passes.

Perphaps the most frequent of all pressure systems that affect the British Isles is the *secondary depression*, a small but vigorous low-pressure system contained within a parent depression. Many secondaries travel across southern England or up the English Channel, where they cause severe storms, whilst the parent depression crosses northern England or Scotland.

Anticyclones and Ridges

More settled weather is associated with *anticyclones*. These are high-pressure areas, where the air is slowly descending. Air that is descending becomes compressed and warmed, so that rain is unlikely to fall. The wind, if any, is light and blows outwards in a clockwise direction. Anticyclones are generally slow moving or may remain stationary for days. In summer they produce clear skies, with calm, sunny weather. There may be early morning mist or dew, and local winds, such as land and sea breezes, may develop. In winter the clear skies and lack of wind allow frost and fog to develop. In the daytime it may be sunny; on the other hand there may be dull weather, with overcast skies.

A similar but short-lived system is the *ridge of*

high pressure, in which the isobars form a blunt V, generally pointing northwards, with depressions to the east and west. It gives a short period of bright, sunny weather, lasting 12 hours or so, before the next depression arrives.

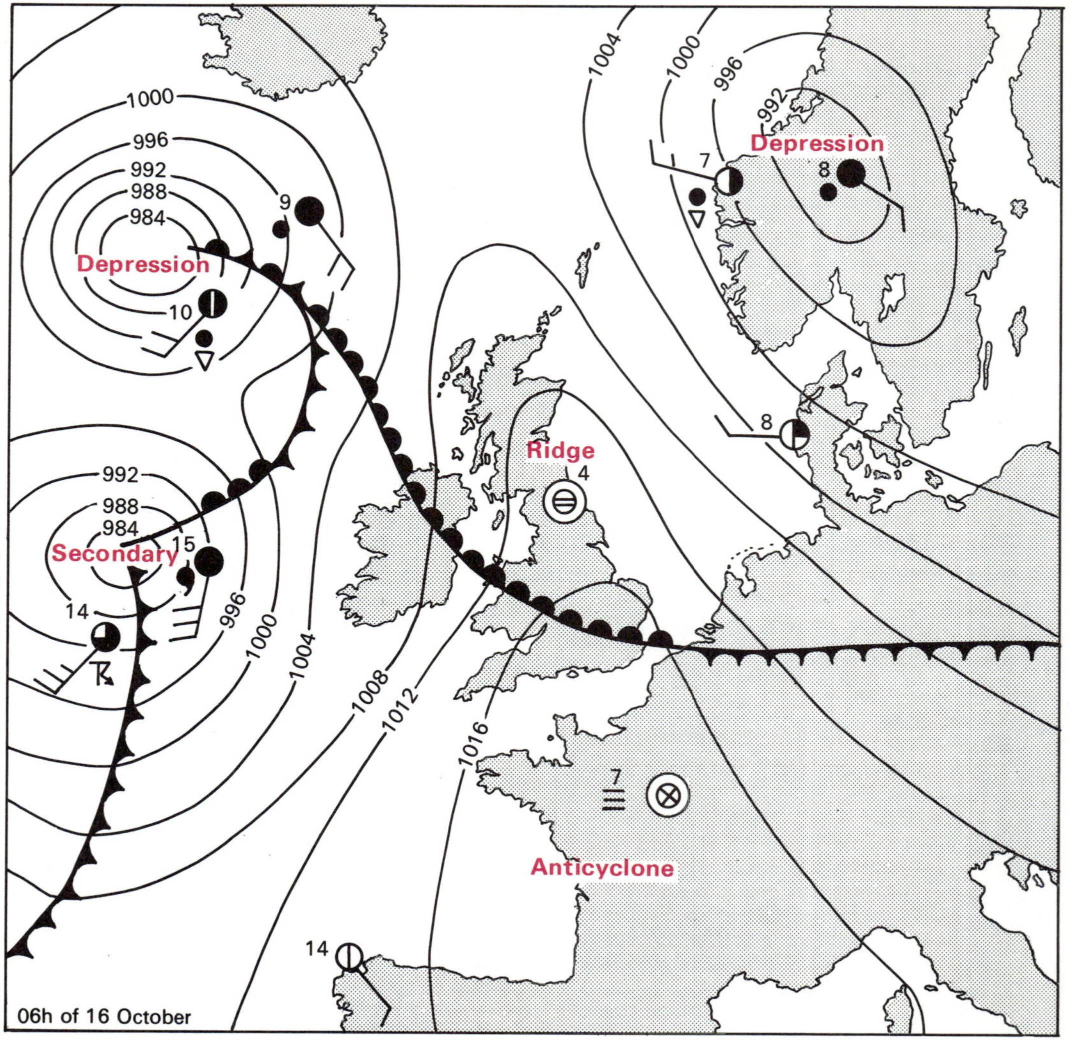

Figure 5.7 A typical weather map

Exercises

1 Study the weather map on the opposite page.
   (*a*) Describe the weather (i) in Northern England and (ii) south of Iceland.
   (*b*) Explain why there are strong winds in the Atlantic to the west of Ireland but calm conditions over France.
   (*c*) How is the weather in Northern England likely to change during the next 24 hours?

2 Study the detailed television weather forecasts for one week.
   (*a*) Note the type of weather system each day and the forecast as it affects your part of the country.
   (*b*) Make a daily note of the actual weather your area experiences and compare the weather with the forecasts.

3 Explain why the weather map on the opposite page shows fog in France.

4 **Your Own Weather Project**
Some equipment will have to be bought but you can make simple versions of other items like rain gauges and Stevenson screens. Garden centres usually sell a range of useful instruments. Many weather elements are observed by eye.

The weather station should be set up in your garden and the readings taken daily at a fixed time. The readings and observations should be entered in a log book and used for working out the climatic data. Daily weather maps which appear on television or in the newspapers should be consulted before you issue a forecast.

5 How are the following elements of the weather measured?
   (*a*) Temperature
   (*b*) Humidity
   (*c*) Pressure
   (*d*) Wind
   (*e*) Cloud type and cover
   (*f*) Precipitation
   (*g*) Visibility
   (*h*) Sunshine

TABLE OF RELATIVE HUMIDITY

(per cent)

Depression of Wet Bulb

		1	2	3	4	5	6	7	8	9	10
Dry Bulb	**40**	93	87	82	76	71	66	61	56	52	47
°C	**35**	93	87	81	75	69	64	58	53	49	44
	30	92	86	79	73	67	61	55	50	44	39
	25	92	84	77	70	63	57	50	44	38	
	20	91	83	74	66	59	51	44	37		
	15	90	80	71	61	52	44	35			
	10	88	76	65	54	44	34				
	8	87	75	63	51	40					
	6	86	73	60	47	35					
	4	85	70	56	42						
	2	84	68	52	37						
	0	82	65	48							
	−2	80	61	42							

Where readings are in between these figures a simple calculation will be necessary.

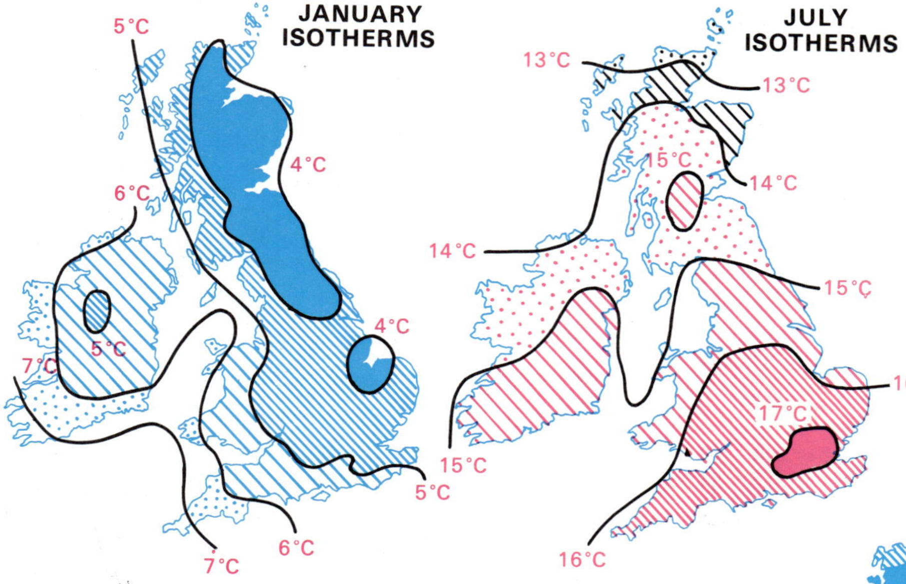

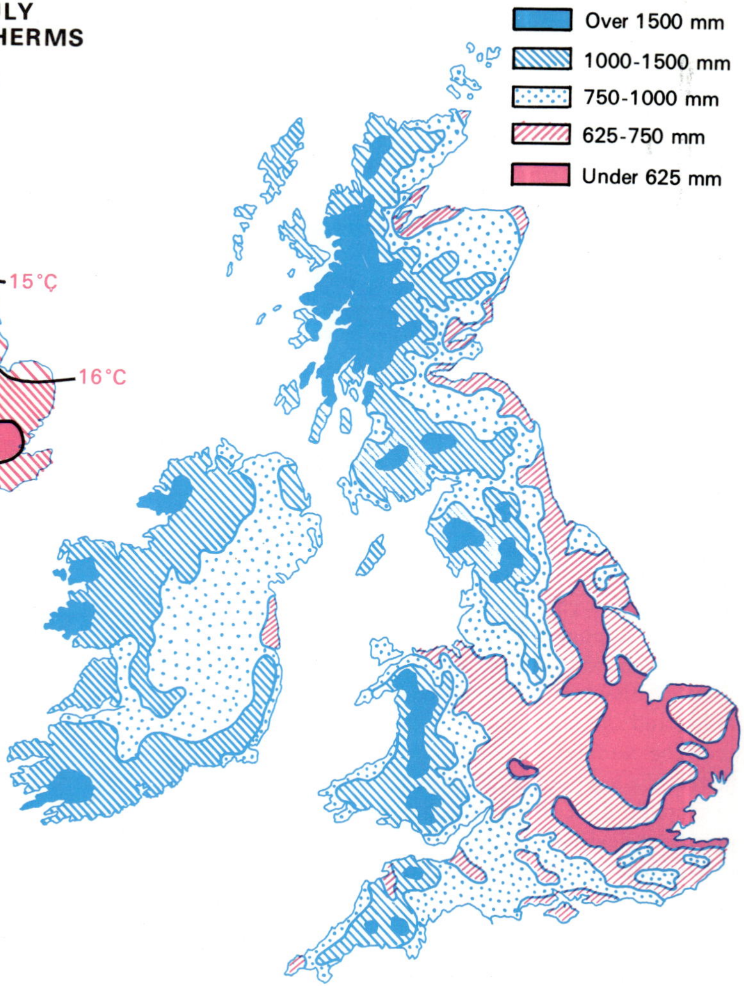

Figure 5.8
January and July isotherms and annual rainfall

Study these maps and then answer the following questions: (*a*) Which are the warmest and coolest parts of the British Isles in July? (*b*) Where is it mildest in winter? (*c*) Why do the July isotherms tend to bend northwards over the land and southwards over the sea, whilst those for January bend southwards over the land and northwards over the sea? (*d*) Which are the wettest and driest parts of the British Isles, and what are the main reasons for the differences?

Pico Viejo, Tenerife (*D Jones*)

PART THREE

The Surface of the Earth

Chapter 6

The Earth's Crust and Earth Movements

Jurassic Limestone, showing strata, joints and bedding planes (*F R Dobson*)

It is generally accepted that the Earth was originally a hot, gaseous mass, and that after it cooled down to a liquid state a solid crust formed on its surface. The age of the Earth is estimated to be about 5000 million years, though the rocks of its crust which are accessible to us were almost all formed during the last 1000 million years. There are three main types of rock, each formed in a different way:

1 Igneous rocks. These originated as molten matter (magma) which cooled and solidified. Lava and granite are two types of igneous rock. Lava cooled quickly on or near the Earth's surface and is generally of a uniform dark colour. Granite cooled slowly beneath the surface, so that the minerals which it contains (chiefly quartz, feldspar and mica) had time to form crystals.

2 Sedimentary rocks. These were formed from material deposited on the sea bed, or occasionally on land. In deep water limestone and chalk have formed from the calcium carbonate left behind by animals which lived in the sea, whilst in shallow water sandstone and clay have formed from material derived from the land. On the land itself coal is the remains of former vegetation which grew in swampy areas.

Sedimentary rocks frequently occur in layers or *strata*, separated by long cracks called *bedding planes*. In most cases the strata were horizontal when laid down, but later earth movements may

A SUMMARY OF GEOLOGICAL HISTORY

Era and Approx. Duration	Period	Characteristic Formations in United Kingdom	Appearance of Life	Mountain Building Movements (Orogenies)
Quaternary 2 Million Years	RECENT	River terraces, raised beaches	Stone Age Man	
	PLEISTOCENE	Ice Age		
Tertiary (Cainozoic) 50 Million Years	PLIOCENE	Norwich Crag	Early Man?	
	MIOCENE			Alpine
	OLIGOCENE	Hampstead Beds		
	EOCENE	London Clay		
Secondary (Mesozoic) 140 Million Years	CRETACEOUS	Chalk, Greensand, Gault Clay	Flowering plants	
	JURASSIC	Jurassic Limestone, Lias, Oxford Clay, Kimmeridge Clay	Birds	
	TRIASSIC	New Red Sandstone	Mammals	
Primary (Palaeozoic) 350 Million Years	PERMIAN	Magnesian Limestone	Conifers	
	CARBONIFEROUS	Coal Measures, Millstone Grit, Carboniferous Limestone	Reptiles Ammonites	Hercynian
	DEVONIAN	Old Red Sandstone	Ferns	
	SILURIAN		Fishes	Caledonian
	ORDOVICIAN	Ancient slates and shales	Trilobites, Brachiopods	
	CAMBRIAN		Gastropods	
Archaean	PRE-CAMBRIAN	Gneiss, Quartzites, Schists, Slates, Igneous Rocks	Sponges, Seaweed	Charnian

have caused them to dip or to form a series of folds. Other cracks are often present, roughly at right angles to the bedding planes, and are known as *joints*. The strata should not be confused with the minute layers (laminae) which occur in some rocks like shale.

3 Metamorphic rocks. These have been altered from their original state by great heat or pressure. In this way limestone has been converted to marble and sandstone to quartzite. Some rocks, including clay, when subjected to great pressure, have been changed to slate. The latter has well-developed lines of cleavage, along which it can be split easily into thin sheets.

EARTH MOVEMENTS

Evidence of Uplift and Subsidence

It is not difficult to find fossil sea shells embedded in limestone and other rocks many hundreds of metres above sea level. This suggests that an uplift of the land relative to sea level has taken place, or alternatively a fall in the level of the oceans. A more recent uplift is indicated by the *raised beaches* in Scotland, Scandinavia and elsewhere. These may be 8 metres or more above the present high-water level, and the beaches often end up against wave-cut cliffs, with caves and other features associated with a coastline.

On the other hand a subsidence of the land is shown by the presence of numerous drowned valleys, known as *estuaries*, round our coasts. These estuaries are the lower courses of the rivers flowing into them, which have been occupied by the

Raised beach in Scotland (*Crown Copyright*)

River estuary in Devon (*Aerofilms*)

What evidence of earth movements is to be found in these photographs? Describe and explain each scene, illustrating your answer with line drawings.

sea. More evidence can be found in the remains of *submerged forests*, consisting of tree stumps and roots in the actual position of growth, together with deposits of peat, visible at low tide in parts of Devon, Cornwall and other coastal areas.

Earthquakes

Earthquakes are a series of rapid vibrations, originating from the slipping or faulting of parts of the Earth's crust at depths varying from 10 to 30 kilometres. On an average about four earthquakes occur each day, but only a few do serious damage. The great majority take place within well-recognised zones which are marked in figure 6.2. The close connection between the distribution of earthquake zones and volcanoes can also be seen.

Severe earthquakes are often associated with coastlines where mountains slope steeply to ocean deeps, for example the east coast of Japan. Here the Tokyo earthquake of 1923 produced considerable local alterations in sea level and numerous fissures. About 250 000 people were killed or injured, mainly as a result of falling masonry and fires following the fracture of gas mains.

Earthquake shocks are recorded on delicate instruments called *seismographs*, which are capable of detecting vibrations due to earthquakes thousands of miles away.

Figure 6.1
Earthquake waves as recorded on a seismograph

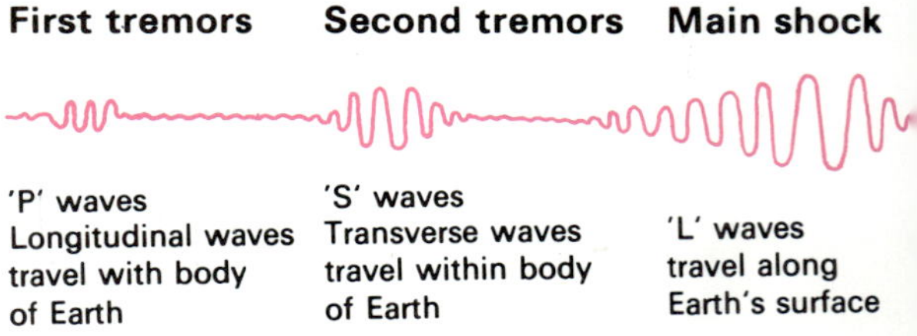

Figure 6.2 Major structural features of the earth

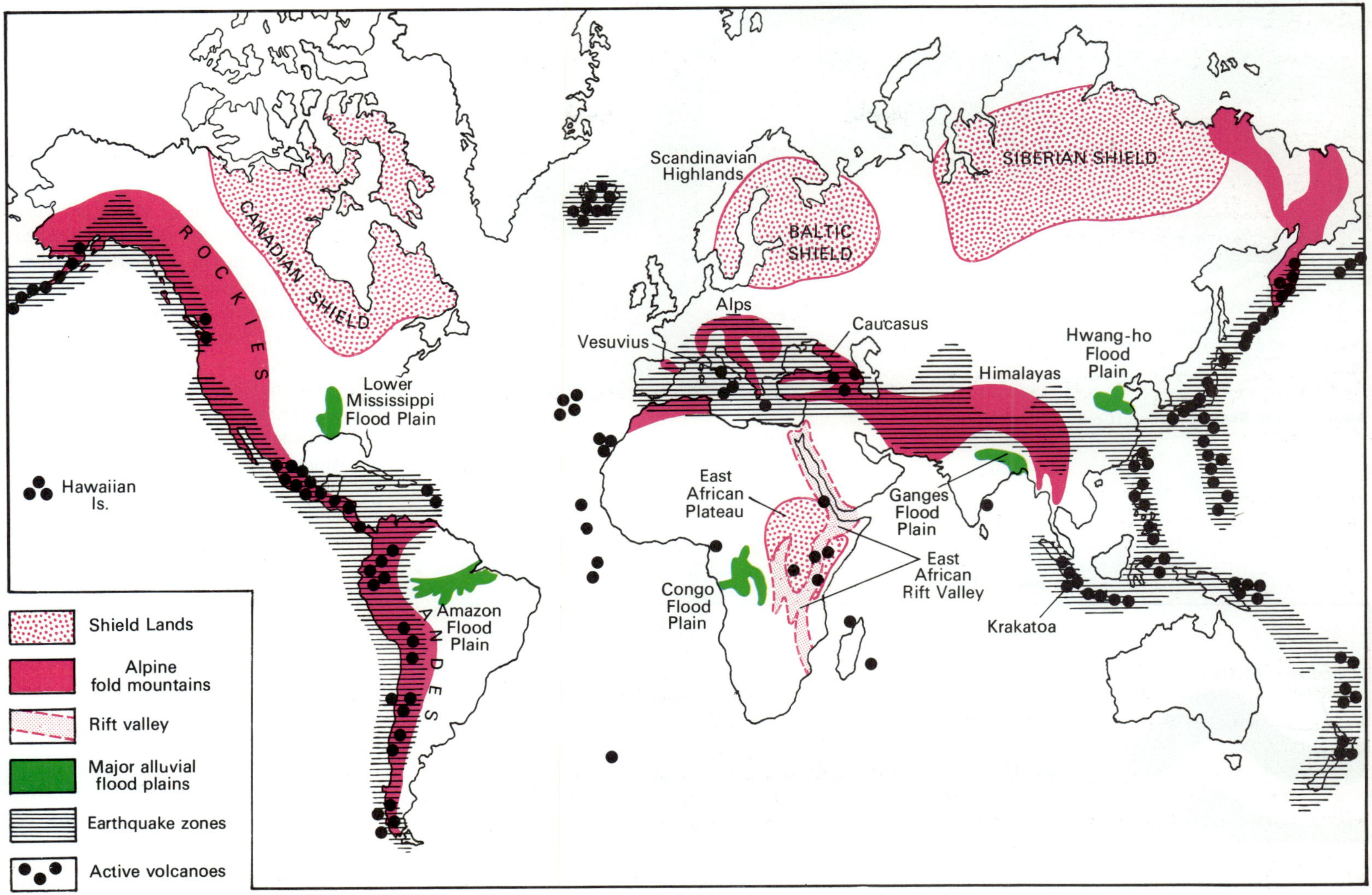

Anticline near Bude, Cornwall (*Crown Copyright*)

Fold Mountains

At certain points in geological history vast disturbances of the Earth's crust have occurred. Prior to these disturbances there were long, quiet periods during which sedimentary rocks, thousands of metres thick, accumulated. These sedimentary deposits were then subjected to enormous pressure by earth movements of old blocks (shields) of hard rock, and forced upwards into a series of folds. Sometimes simple upfolds (*anticlines*) and downfolds (*synclines*) were formed, as in the Jura Mountains. In other cases the folds were pushed over on one side, producing *overfolds*, whilst in some of the highest fold mountains, like the Alps, great folded masses called *nappes* have been thrust or squeezed forward for many kilometres from their original positions.

These mountain-building movements, known as *orogenies*, have taken place on a number of occasions during geological history. The most recent one was the Alpine orogeny, which created the Alps themselves, and many other great mountain ranges, including the Himalayas, Rockies, Andes and Caucasus.

Figure 6.3 Folds

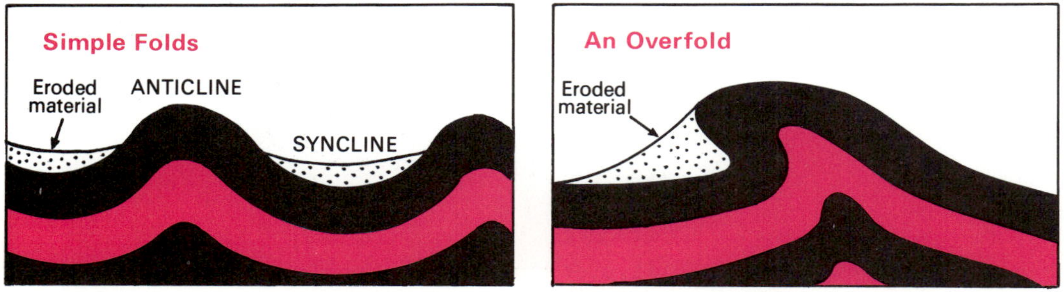

Shield Areas

Over millions of years, land surfaces may be worn down by erosion until they become almost level plains. There may be only a few remnants of the original mountains. Areas like this generally consist of very old rocks and are called *shields*. They are some of the most stable parts of the earth's surface.

Faulting

Both pressure and tension in the Earth's crust can produce *faults*. These are fractures along which movement has taken place. One side may have slipped downwards, leaving the upstanding side as a *fault scarp*. The Pennine Fault, in the northern Pennines, west of Cross Fell, has produced such a scarp, overlooking the River Eden.

Where parallel faults have occurred the land between may subside giving rise to a long, narrow depression called a *rift valley*.

It may also happen that the land between faults may be forced upwards to form a plateau or *horst*. The Vosges and Black Forest, separated by the Rhine Rift Valley, were formed in this way.

IGNEOUS ACTIVITY

At certain points of weakness in the Earth's crust, molten lava may force its way upwards, reaching the surface through a 'pipe', at the upper end of which is a circular *crater*. Round this crater solidified lava and other materials accumulate to form a *volcano*.

Lava Volcanoes

A volcanic eruption may produce a succession of lava flows down the volcano's sides. Some lava, called *acid lava*, is rich in silica. Because of this it flows slowly and solidifies quickly, giving rise to a steep-sided cone. After a time the pipe becomes plugged with solidified lava and the volcano is said to be dormant. The dormant period may last for a considerable time and may be followed by a violent eruption. Basic lava, on the other hand, contains less silica and is more fluid. It thus travels great distances before it solidifies, so that a gently sloping volcano is produced, which erupts frequently but without great violence.

An example of an acid lava volcano is Krakatoa, an island in the East Indies. Its greatest eruption was in 1883 when two-thirds of the island disappeared in a tremendous series of explosions. Vast quantities of glowing pumice and ashes, accompanied by gases and steam, were hurled high into the air. The eruption produced 'tidal' waves which drowned 36 000 people, and a huge island-rimmed crater, over six kilometres across, was left. Such an enormous crater is called a *caldera*.

Rhine Rift Valley, were formed in this way.

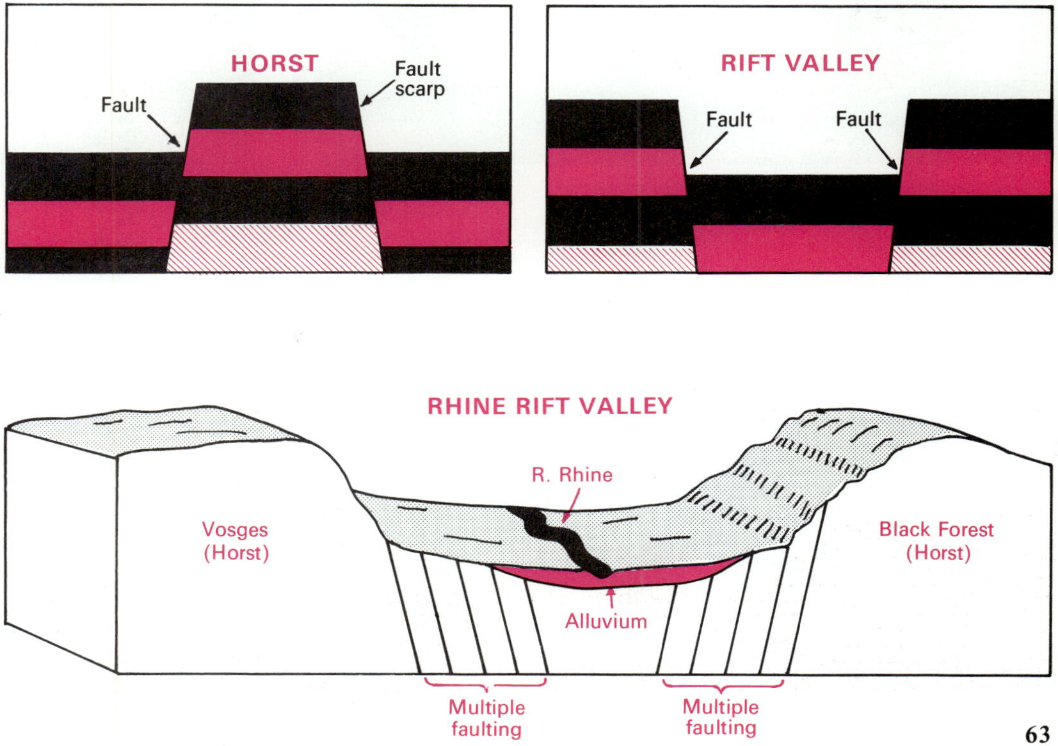

Figure 6.4 Features of a faulted landscape

Etna in eruption, 1971, north-east crater and lava flow (*S A R Usmate and Sicilia Folklore, Messina, Italy*)

Etna is Europe's greatest volcano. Its history is one of mild, continuous activity in the central craters, with minor vents opening up on the flanks. These photographs are of the flank eruptions of 1971, when there were powerful explosions followed by lava flows, which caused considerable damage but no loss of life.

Figure 6.5 Section through a composite volcano

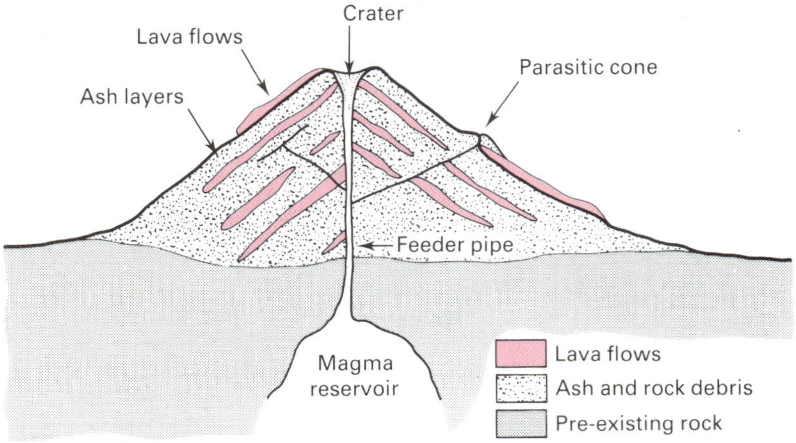

Figure 6.6 Igneous intrusions

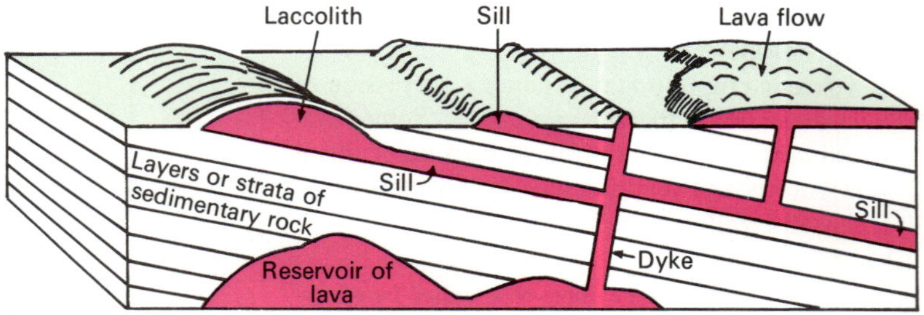

Kilauea is a gently sloping volcano of the basic lava type in the Hawaiian Islands. Its crater is normally occupied by a lake of seething lava, whose level rises from time to time, so as to produce new lava flows. Unlike Krakatoa, which remains dormant for long periods, Kilauea is in a state of almost constant eruption, in a quiet way, and is usually quite safe to approach.

Ash or Cinder Volcanoes

Some smaller volcanoes consist of fragmental material, including blocks of lava of all sizes, volcanic bombs (rounded masses of lava formed as a result of whirling and cooling in flight), and finer-grained volcanic ash. All these were thrown upwards by successive eruptions and on falling to the ground collected round the crater, to give a conical shape.

Composite Volcanoes

Most large volcanoes are built up of both lava and ash. Lava streams may also emerge from fissures in the side of such volcanoes, and there may be subsidiary cones. Vesuvius, which is of this type, had a gigantic basin-shaped crater in the time of the Romans, but in A D 70 it erupted violently, destroying half of the crater wall and burying the towns of Pompeii and Herculaneum under volcanic ash and mud. Since then there have been many eruptions and a new cone, with a crater at its summit, has arisen within the broken wall of the old crater, the surviving side of which is called Monte Somma. A number of cinder cones also lie within the old crater.

The Whin Sill, with Hadrian's Wall (*D Jones*)

Fissure Eruptions

Molten rock may well up along large fissures and spread out to form lava sheets. The Plateau of Antrim, in Northern Ireland, and parts of the western isles of Scotland were formed in this way. The lava is of a basic type, known as basalt. As it cooled it sometimes developed a columnar structure, due to contraction. These columns are usually six-sided and may be seen at the Giant's Causeway, in Northern Ireland, and at Fingal's Cave, on the isle of Staffa.

Igneous Intrusions

Magma may penetrate other rocks to form igneous intrusions. These take the form of *sills* where they occupy the bedding planes and *dykes* where they occupy the joints. Good examples are the Whin Sill in Northumberland, whose defensive advantages were recognised by the Romans when they built Hadrians's Wall along it, and the Cleveland Dyke in the North York Moors.

Another type of intrusion is called a *laccolith*. It is a giant bubble-shaped mass of magma which presses up the overlying rocks into a dome. The granite bosses of Dartmoor and Bodmin Moor were formed in this way, the surface rocks having been worn away so as to reveal the granite.

left **Edinburgh Castle rock, a former volcano;** *right* **lava plateau on the Isle of Staffa** (*Aerofilms*) Note the columnar jointing of the basalt.

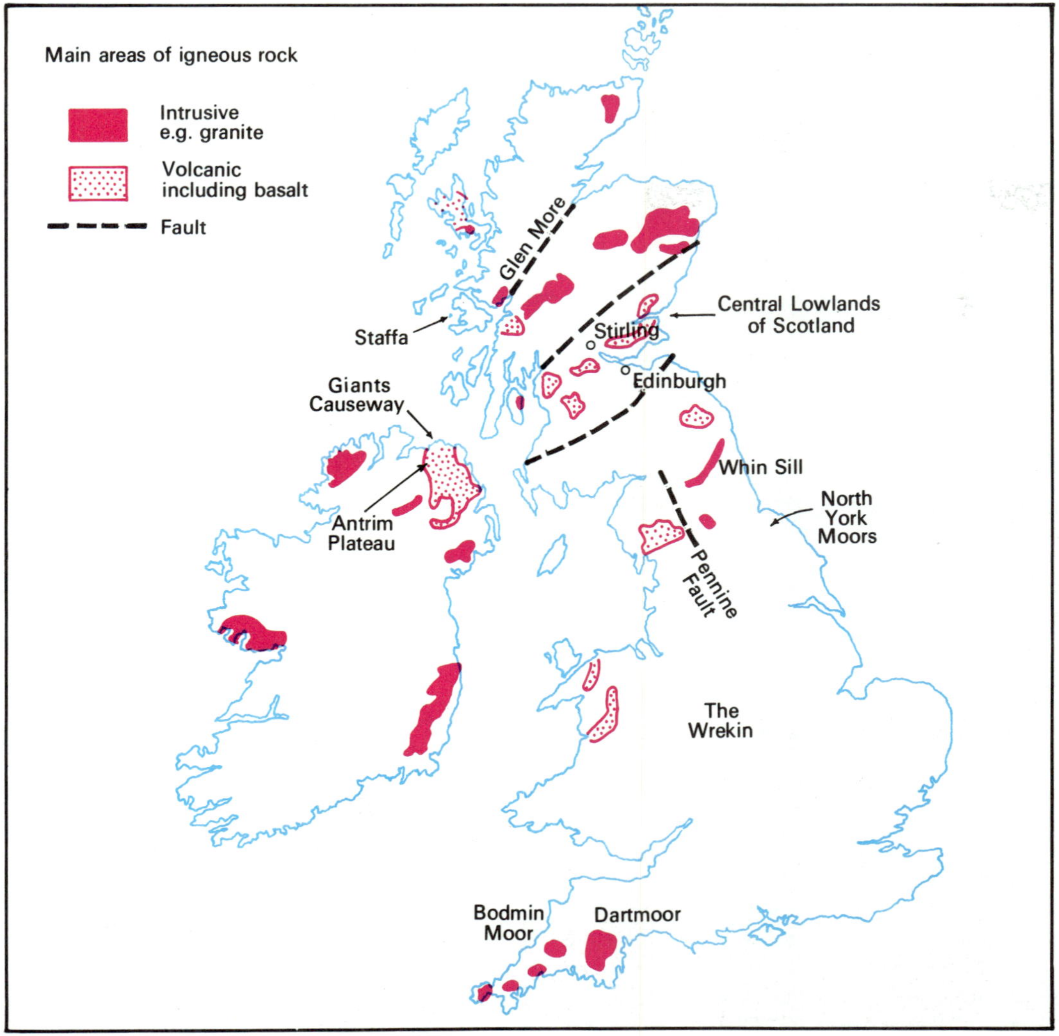

Hot springs and *geysers* also point to igneous activity. Geysers occur where ground water comes into contact with hot rocks. There must be a narrow pipe leading upwards to the surface. The water at the bottom is under pressure, which raises its boiling point. When it does boil some of the water is forced out of the pipe, and the release of pressure causes practically the whole contents of the pipe to be ejected in a fountain of steam and boiling water. Geysers occur in the Yellowstone National Park (USA), Iceland and New Zealand.

Evidence of Past Igneous Activity

Although there are no active or dormant volcanoes in the British Isles there is plenty of evidence of past activity. A glance at a geological map shows the widespread distribution of igneous rocks. Actual volcanoes, however, are difficult to trace, though the Wrekin in Shropshire, and the hills on which Edinburgh and Stirling Castles stand, are almost certainly the remains of former volcanoes. In Europe there are many extinct volcanoes, often with circular crater lakes, especially in the Auvergne region of France and the Eifel in Germany.

Figure 6.7
Evidence of past igneous activity in the British Isles

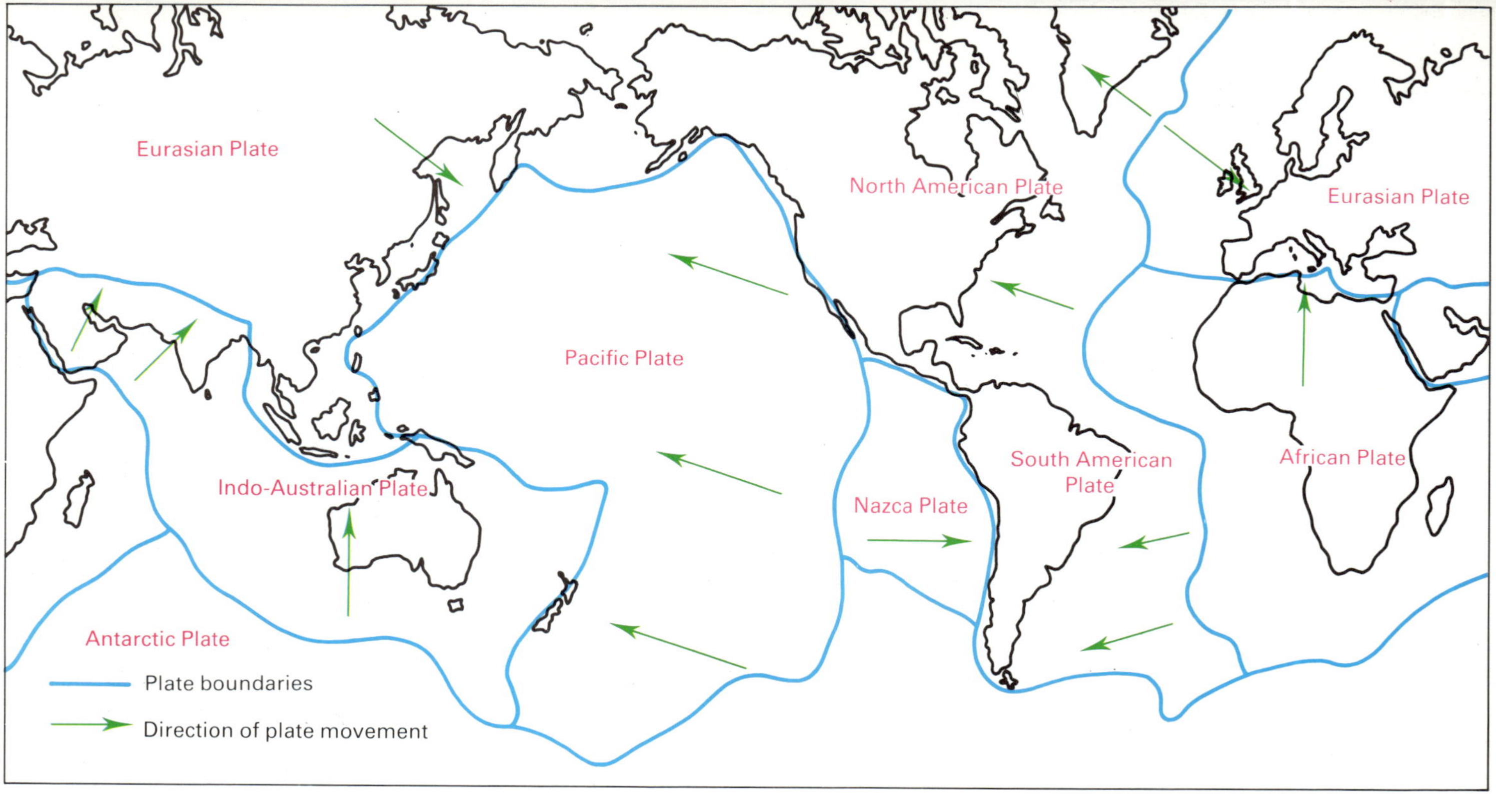

Figure 6.8 Movement of the Earth's crust

Volcanoes, Earthquakes and Continental Drift

Volcanoes and earthquakes do not occur in a random pattern. Figure 6.2 showed a close link between the distribution of earthquakes and volcanoes. Both occur along lines of weakness which mark the edges of the big sections or *plates* which form the Earth's crust. It is at the *plate boundaries* that most volcanic and earthquake activity is found. One example is the 'ring of fire' around the Pacific Ocean where the volcanoes and eathquakes mark the edge of the Pacific plate

There are two main kinds of plate boundary. There are those where the plates are moving apart and those where the plates are moving towards each other. For example, the island of Iceland is the tip of a huge volcano on the Mid-Atlantic Ridge which marks a major plate boundary along which the Atlantic Ocean is widening. New rock wells up along this boundary from below the crust. By contrast, the volcanoes of Japan and the Phillipines occur where the Pacific plate is being forced down beneath the Eurasian plate. The rock material of the Pacific plate is melted by the heat, and some of this molten material forces its way to the surface to form volcanoes.

Movement of the plates against each other creates shock waves or earthquakes, which were described earlier in the chapter. Plate movement also explains the formation of fold mountains, with sediments in the oceans being crumpled up as plates move towards each other (figure 6.8).

'Old Faithful' geyser (*U S Information Service*)
This geyser in Yellowstone National Park shoots up a 54 metre stream of scalding water every 65 minutes. The display lasts about four minutes.

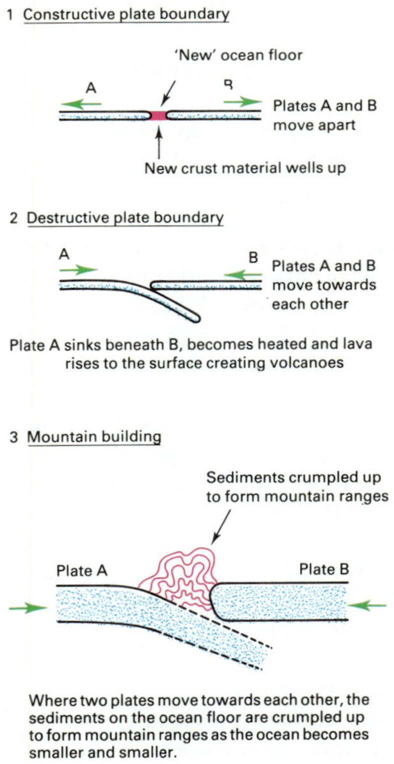

Figure 6.9 Plate boundaries, new ocean floor and fold mountains

Exercises

1 (*a*) Name one recent natural disaster.
   (*b*) Describe briefly what happened.
   (*c*) What were the effects of the event?
   (*d*) What measures can be taken to reduce the effects of such events?

2 Study figures 6.2, 6.8 and 6.9. Explain why volcanoes and earthquakes occur in Southern Italy.

3 Describe the patterns of earthquake zones and volcanic zones. Explain why they are so similar.

4 Make simple, clear, labelled drawings to show the following:
   (*a*) A cliff face composed of a horizontal sedimentary rocks made up of thick beds with clearly marked joints.
   (*b*) A cliff face in sedimentary rocks folded into an anticline and a syncline. The sedimentary rocks are made up of alternate layers of dark and light coloured rocks.

5 Refer to figure 6.7.
   (*a*) In which of the following parts of the British Isles do you find intrusive rocks?
      (i) East Anglia
      (ii) Central Lowlands of Scotland
      (iii) South West England
      (iv) North Wales

6 **Making a Geological Collection**
The best way to learn about rocks is to handle and examine them yourself. Specimens are easy to find, in quarries, clay pits, along the sea coast and elsewhere. A note should be made of the exact location. The rocks should later be examined carefully, preferably under a magnifying glass, and the colour, hardness, texture, fossils and other characteristics noted. Reference should be made to a geological map of the area, so that the type and age of each rock can be found. For the identification of fossils a geological handbook should be consulted, or a visit made to the local museum. Finally each rock should be carefully labelled, with information about its type, age, any fossils in it and where it was found.

Chapter 7

Earth Sculpture in Humid Regions

WEATHERING

The rocks which make up the surface of the earth are constantly under attack by *physical* and *chemical weathering.*

Physical Weathering

One kind of physical weathering occurs when water derived from rain or melting snow seeps into cracks in exposed rocks. In high mountains the water freezes. This causes it to expand, and when the process of freezing and thawing is repeated, it eventually breaks up the rock. The loosened fragments fall downhill, the large boulders often ending up in the valleys, whilst the smaller fragments come to rest on the lower slopes where they form *screes.*

Physical weathering is also carried on by plants, especially trees, which frequently extend their roots into cracks in the bedrock and exert a surprising force to break apart the rock.

Chemical Weathering

Chemical weathering also causes the disintegration of rock. Rain water absorbs carbon dioxide from the air and becomes a weak carbonic acid. Rocks such as limestone and chalk, which are composed of calcium carbonate, are highly soluble, and the result of weathering is water very rich in calcium bicarbonate, or hard water. Other rocks, too, are broken down in the presence of carbonic acid, although there is usually a considerable amount of insoluble residue. The best-known example is the way in which the feldspar and mica in granite break down into clay, leaving the other mineral, quartz, as a sand within the clay.

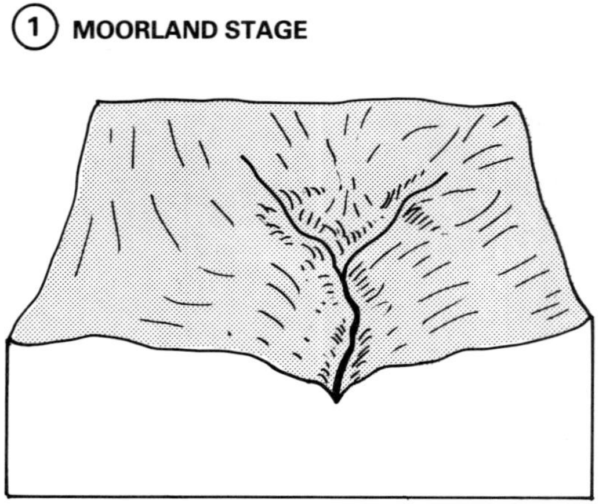

Figure 7.1 The stages of a river

RIVERS

In areas of *impermeable* rock (rock through which water cannot pass), rain water flows on the surface in small rivulets which join up and cut small channels by which the water is carried away. If such an area is generally level, it may be covered by thick peat deposits which act as a sponge for rain water. As a result these rivulets only dry up after exceptionally dry periods. The *source* or starting point of a stream may be an area of peat as described, or it may be a spring, a lake or in high mountains like the Alps it may be the melting point of a glacier.

The typical river in its upper valley course has a steep gradient and occupies a narrow V-shaped valley. This V-shape is a result of (*a*) the process of vertical erosion, by which is meant the lowering of

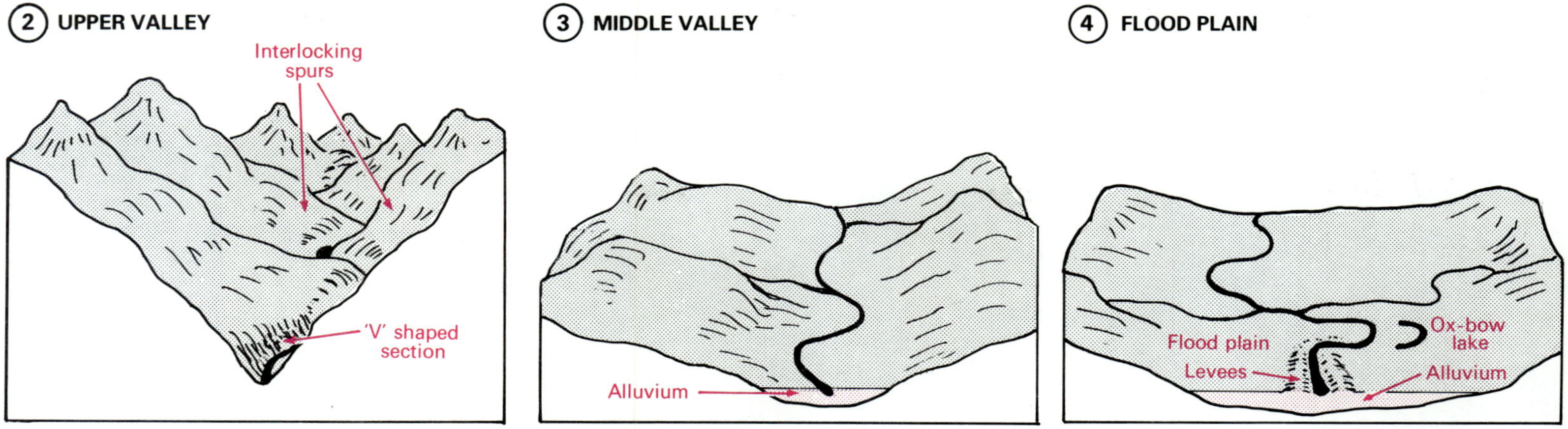

the river bed by the scraping action of sand, pebbles and boulders and, in particular, by the creation of *potholes* in the channel bed, and (*b*) the wearing down of the valley sides by weathering. The pebbles and boulders mentioned under (*a*) are themselves rounded and worn down by being knocked against each other and against the rocky bed.

At first, most streams follow a fairly straight course but after some distance they begin to take a winding path. The result is a series of interlocking spurs.

Further downstream the gradient of the valley becomes more gentle. There follows a change from vertical to *lateral erosion*, so that the valley floor is widened by erosion on the outside of the bends where the current is faster. The outside bank of the river channel is cut back, with a low cliff edge. On the inner bank, where the current is slow, material is deposited to form a gently sloping beach. As time goes on, the bends or meanders gradually cut a stretch of flat land. Since downward erosion is now very slow, weathering of the valley sides and movement down the slope of weathered material produces gentler and more rounded sides to the valley. When the river floods, it deposits a thin layer of silt. The material deposited in this way is called *alluvium*, giving the name *alluvial* or *flood plain*.

The meanders of the river develop broad loops which slowly migrate downstream. In times of flood the narrower necks of meanders are sometimes broken through, leading to the abandonment of the meanders and the formation of *ox-bow lakes*.

Where the alluvial plain is broad and low-lying, the river forms numerous meanders and ox-bow lakes. Drainage ditches are often dug to carry surplus water into the river, whose banks may be

Figure 7.2 Stages in the development of an ox-bow lake

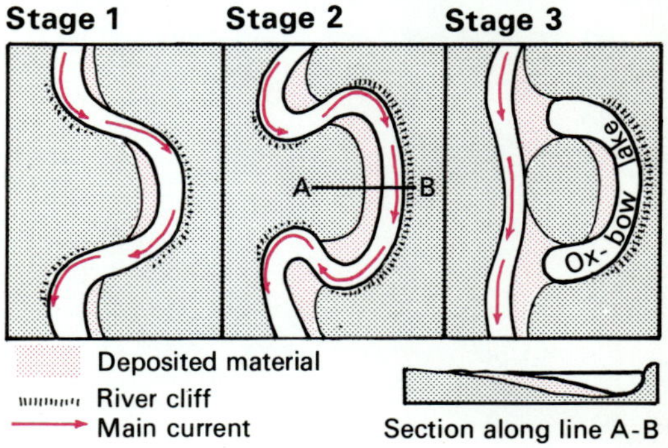

raised artificially to protect the plain from being flooded. Such raised banks are termed *levees*, a name originating from the River Mississippi.

Waterfalls, Rapids and Gorges

As has already been explained, the gradient of a normal river becomes less and less steep as it flows downstream from its source to the sea. But where hard rock lies across a river's course it reduces the gradient above it, and causes this to increase again farther down-stream. As the river crosses the hard rock there may be rapids, and if the hard rock overlies softer rock there will probably be a waterfall at the lower end. The rush of water at the base

The Dolomites; upper valley stream; middle valley river
(*Photos E G Bowley ARPS, F R Dobson*)

of such a fall wears back the softer rock until the hard rock above it is unsupported. Large fragments break away and the fall slowly recedes upstream, leaving a gorge below. The Niagara Falls, which lie on the Niagara River between Lakes Erie and Ontario, in North America, occur where a bed of hard limestone overlies softer shales and sandstones. The falls are about 50 metres high and there is a gorge 11 kilometres long below them. It is known that this gorge has formed since the end of the Ice Age, about 15 000 years ago, and from this it can be calculated that the average rate of recession of the falls has been approximately 75 cm per year. A well-known British waterfall, on a much smaller

Niagara Falls (*National Film Board of Canada*)
Draw a sketch map of the area shown in the photograph, marking with thick lines the position of the Canadian Horse Shoe and the United States Falls, and printing the word GORGE in the correct place. Why have the waterfalls developed here, and how was the gorge formed?

Figure 7.3 Niagara Falls—simplified section

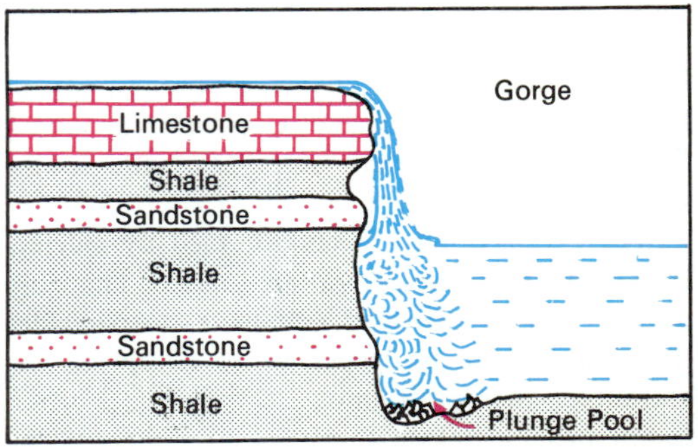

scale, is High Force, in Teesdale, which occurs where the River Tees flows over the Whin Sill on to softer limestone and shales which underlie it.

Gorges are usually produced by a powerful river wearing downwards through hard rock, whilst the valley sides remain nearly vertical because of their high resistance to weathering. An example is the Clifton Gorge near Bristol where the River Avon has cut through an outcrop of Carboniferous Limestone.

In addition, gorges are common in the upper reaches of valleys where the valley gradient is steep, so that vertical erosion takes place at a much more rapid rate than the weathering of the valley sides.

Jordan Rift Valley, showing a river in its flood plain stage, with meanders and ox-bow lakes (*Aerofilms*)

Deltas

Where a river, heavily laden with silt, reaches the sea its flow slackens and much of its load is deposited. In the absence of powerful currents or tidal movement the silt accumulates to form a fan-shaped delta. Sand or dust, blown by the wind, and the growth of vegetation, may raise its level slightly above sea level. On reaching the delta the river divides into a number of channels, called *distributaries*. Changes of course are frequent and serious flooding may be caused.

Where there are currents the seaward end of a delta becomes smoothed, giving it an *arcuate* (arclike) shape. Sand or mud spits may be built up by the currents, with shallow lagoons behind them. In time these lagoons silt up and more spits are formed farther out to sea, so that the delta grows in size. The Nile delta (figure 7.4) is of this type.

Some other deltas are *lobate* in shape. Banks of sand or mud, bordering each distributary, jut out to sea like fingers. Such deltas occur where rivers flow into relatively quiet seas, as in the case of the Mississippi (figure 7.4).

There are no large deltas in the British Isles because of the action of waves, tides and currents off our coasts, but some small deltas have formed where streams flow into lakes, for example where the River Dee flows into Lake Bala, north Wales.

Estuaries

Many rivers as they reach the sea flow into drowned valleys (estuaries), whose formation was explained in Chapter 6. Estuaries provide sheltered harbours, and many of our largest British ports have grown up within them, for example Liverpool on the Mersey estuary. A disadvantage is the accumulation of sand or silt brought down by the rivers. Although the tides may remove some of this, dredging is generally necessary on the approach to ports.

In areas of high land the drowned valleys are called *rias*. These are narrower than the lowland estuaries, have steeper sides and numerous branches where the tributaries have also been drowned. Rias occur in Devon and Cornwall (including Plymouth Sound), south-west Wales and south-west Ireland.

Where valleys have been occupied by glaciers before being drowned, *fiords* are formed. These are usually longer, narrower and have steeper sides than rias. There is deeper water except at the entrance, where a shallow 'threshold' is often found, probably at the former melting point of the glacier. The branch channels generally join the main fiord roughly at right angles. Typical fiord coastlines are found in Norway and north-west Scotland.

River Basins and River Flow

The area drained by a river and its tributaries is called a *river basin* or *drainage basin*. The movement of water that falls as rain or snow through the river basin to the sea brings about the erosion, transport and deposition of rock material that produces the landforms described earlier in the chapter.

The hydrological cycle is influenced at the local level by the features of the river basin. Water can follow several different paths to the river which eventually takes it away to the sea. Figure 7.6 shows these paths, but their importance depends on circumstances. After a period of prolonged rain, the soil would be *saturated* or full to capacity with water, so rain water would be moved by surface flow. Where rocks are impermeable, especially if the soil is thin and unable to hold much water, surface flow would happen quickly. The same applies where slopes are steep. If there is very heavy rainfall, coming down more quickly than it can soak into the soil, surface run-off is again likely. The absence of vegetation also increases the rate at which water reaches the surface, whereas abundant vegetation acts to *intercept* water. The density of the drainage or river network also affects the movement of water, since a dense network drains an area more quickly than a sparse one.

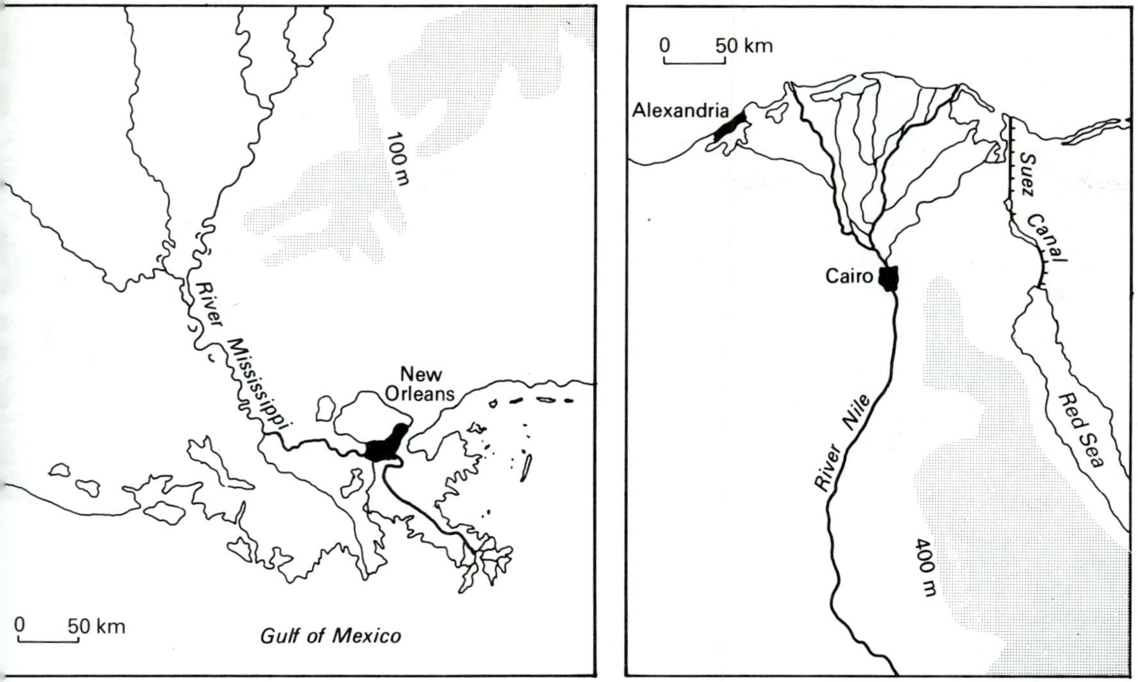

Figure 7.4 Mississippi delta—lobate type; Nile delta—arcuate type

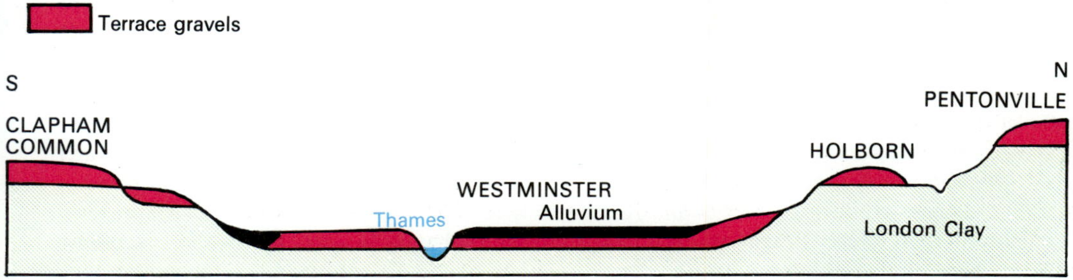

Figure 7.5 River terraces of the Thames at London

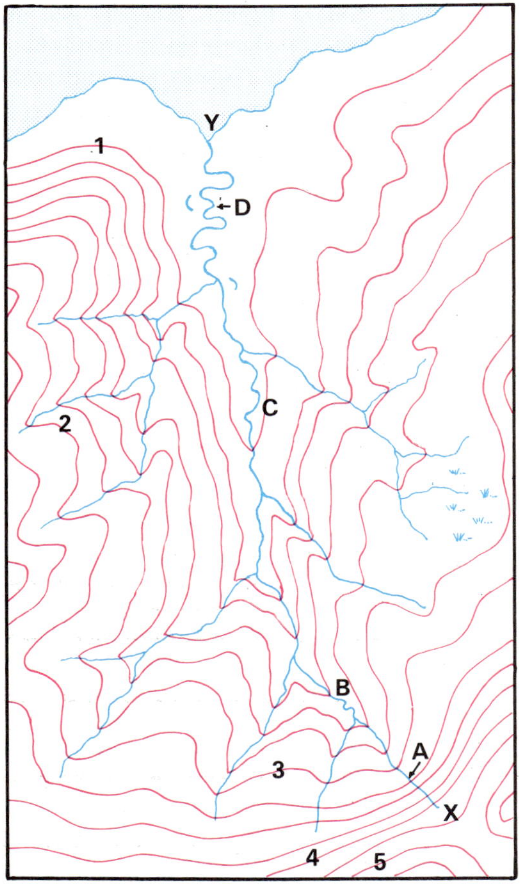

Figure 7.6 A river basin

Calculate the heights at 1, 2, 3, 4 and 5 assuming the contour interval is 50 metres. Draw a profile of the river from its source at X to its mouth at Y. Describe the character of the river and its valley at A, B, C and D.

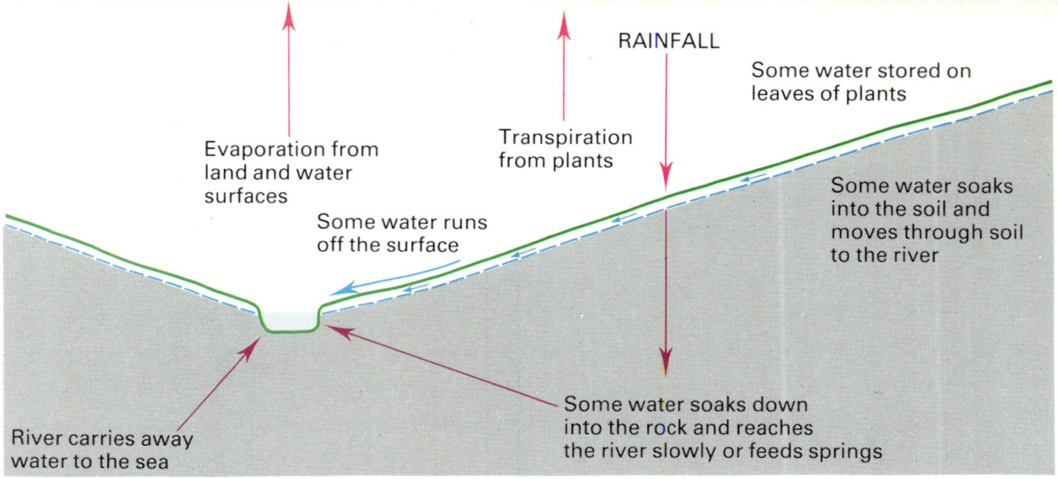

Figure 7.7 The hydrological cycle in a river basin

If all the circumstances described above coincided, it is easy to see why flooding of low-lying areas would be likely to follow.

River Terraces and Incised Meanders

An uplift of the land also has a marked effect. Rivers commence to flow swiftly again, owing to an increase of gradient. Under these conditions a river may lower its bed but retain its meanders, producing *incised meanders*. Some fine gorge-like incised meanders may be seen along the river Wye between Ross and Chepstow, and along the Dee at Llangollen.

Another result of uplift is the formation of *river terraces*. A river may lower its bed and then make a new alluvial plain within, but at a lower level than the original one. Remnants of the old plain may be left as terraces. This whole process may be repeated a number of times, producing a series of river terraces at different levels. Fragments of such terraces have been traced along many British rivers; for example, there is evidence of three different terraces along the Thames valley. Those in the London area are shown in figure 7.5.

UNDERGROUND WATER

In many areas the rocks are *pervious*, and allow water to pass through them. These rocks include limestone, which has deep joints down which the water can penetrate. Others like chalk and some types of sandstone are of a *permeable* or *porous* type and absorb water into the body of the rock itself.

When rain falls on such areas it goes underground, leaving the surface dry and streamless. Eventually it reaches an impervious layer and the downward movement is checked, causing the water to accumulate in the joints, crevices and pores of the rock. The upper level of this ground water is known as the *water-table*. Where the water-table reaches the surface, for example at the foot of chalk

or limestone hills, it gives rise to springs. Elsewhere bore-holes are often sunk to below the water-table, so that water can be pumped to the surface.

Features of Limestone Scenery

The most characteristic form of limestone in the British Isles is the Carboniferous Limestone, a hard greyish rock which forms high land in the Pennines, Mendips and in Wales. Rainwater penetrates its deep joints, some of which are widened by solution to become *swallow holes*, down which streams disappear. The water then follows underground channels, making use of the joints and bedding planes, and dissolving out numerous *caverns*. In these caverns a fantastic assortment of *stalactites* hang like icicles from the roof, and similar but blunter *stalagmites* rise from the floor. Both are caused by the precipitation of calcium carbonate by water dripping from above. The streams generally reappear at the foot of the hills.

◀**The Postojna Cavern, Yugoslavia,** showing stalactites and stalagmites (*Photo F R Dobson*)

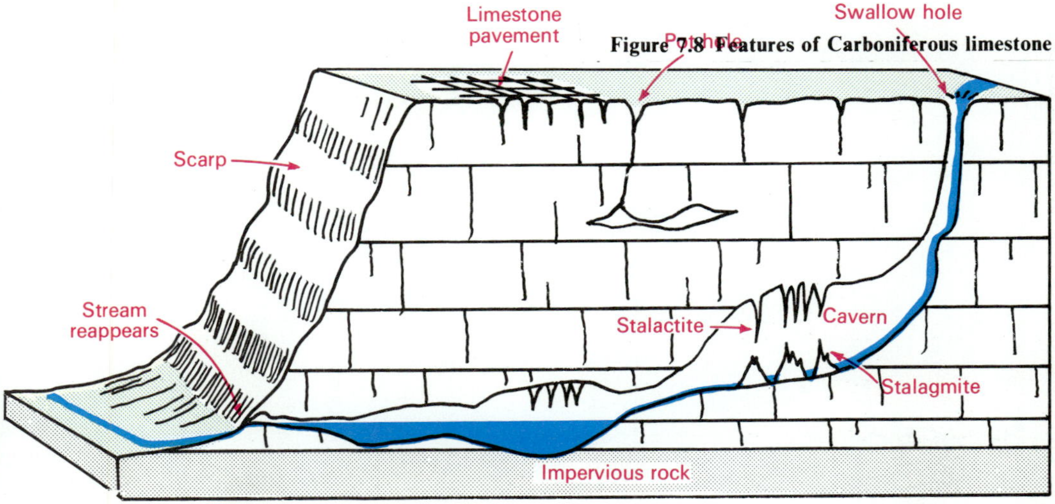

Figure 7.8 Features of Carboniferous limestone

◀ **Roundway Hill near Devizes** (*Aerofilms*)

Malham (*Crown Copyright*)

Study the photographs of Roundway Hill and Malham, and then draw two profiles, (*a*) from the Roundway Hill photograph, to include the lowland on the left, the scarp face and the back slope on the right, and (*b*) from the Malham photograph, to show the almost vertical scarp slope in the middle distance, with the land above and below it.

What types of rock form the uplands in the two photographs, and what connection is there between the type of rock and the scenery?

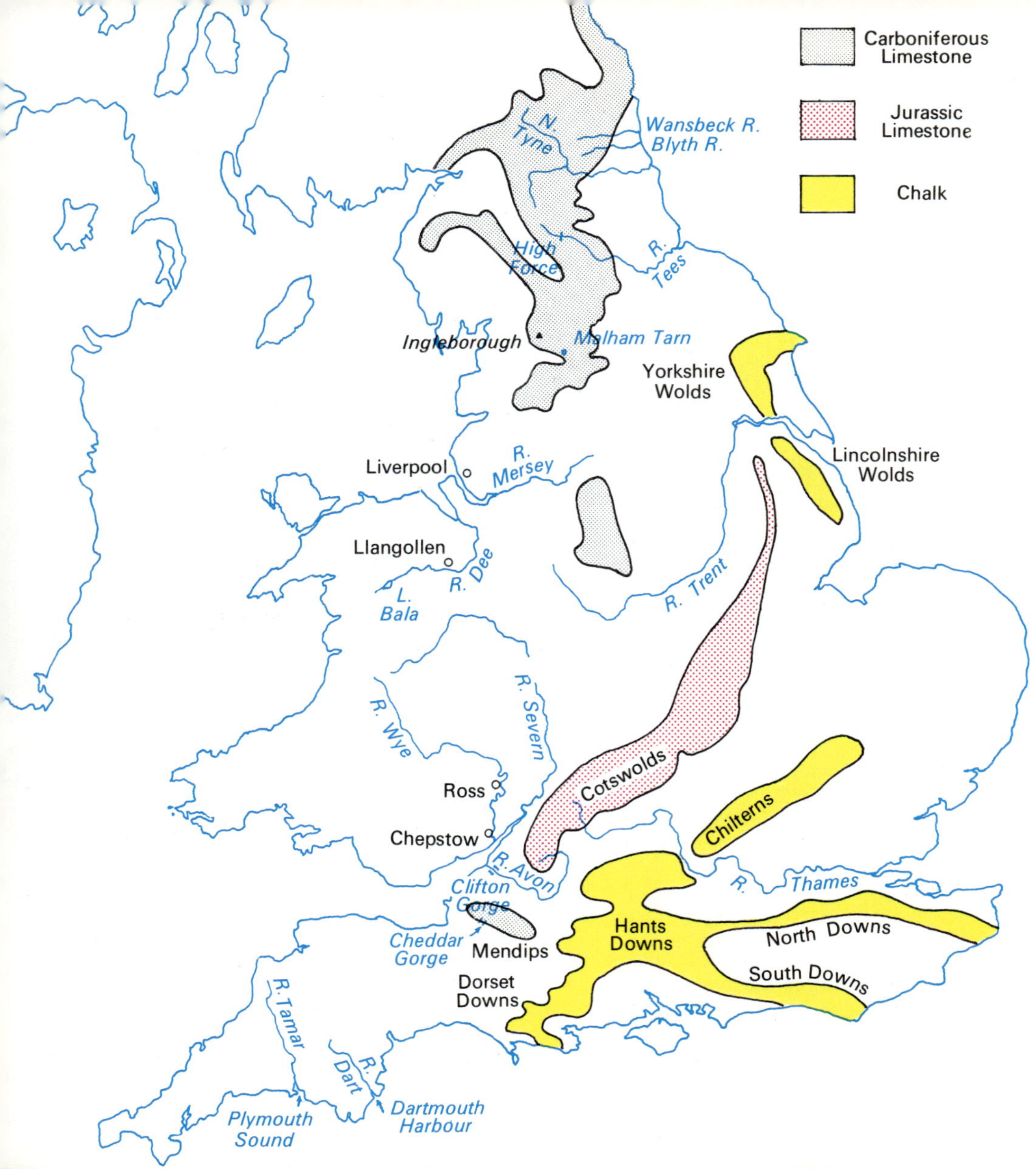

A typical limestone area is around Ingleborough and Malham in West Yorkshire. The Malham area is shown in figure 7.9. From a small lake called Malham Tarn, which lies on impermeable slates, a stream flows on to the limestone and goes underground down a swallow hole. Much of the limestone has a thin soil which is covered with short grass, but in places the bare rock outcrops to form an extensive pavement with slabs of limestone, known locally as *clints*, separated by enlarged joints, called *grikes*. Occasionally potholes allow access to underground passages and caverns. The limestone ends in the south in an abrupt scarp and at Malham Cove a stream emerges from the base of the scarp, flowing across a lowland formed of impermeable Millstone Grit. Near by is a deep, narrow gorge known as Goredale, which has resulted from the collapse of the roofs of a series of caverns, through which flowed an underground river. Such gorges are a common feature of limestone country and include the famous Cheddar Gorge, in the Mendips.

Figure 7.10 Sketch map of the Malham area ▶

Figure 7.9 Limestone and chalk areas in England and river systems referred to in the text

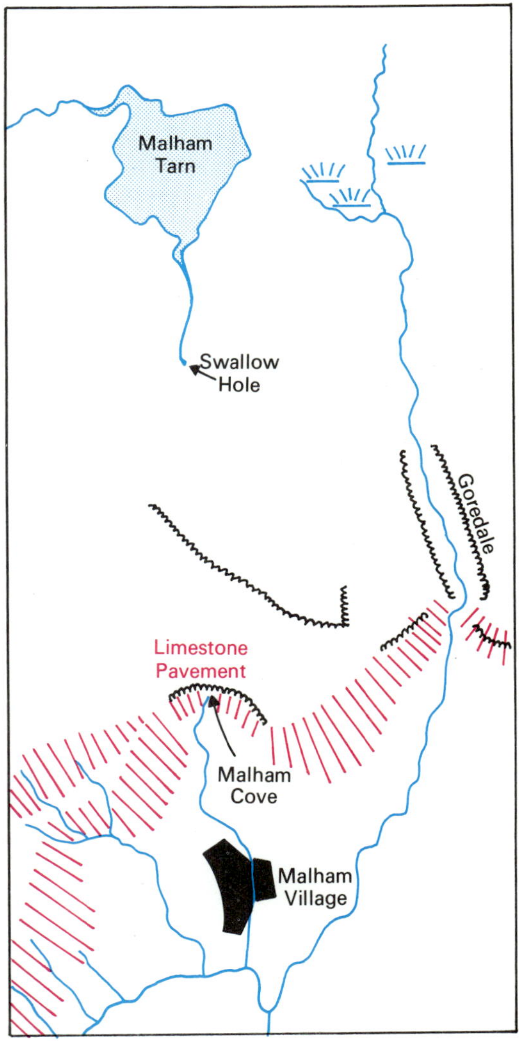

Features of Chalk Scenery

Chalk is a soft, whitish form of limestone, without deep joints. It is a porous rock, so that rainwater soaks into the mass of the rock, rather than down individual joints. Chalk hills occur in the east and south of England, and are known as Wolds in Yorkshire and Lincolnshire, and Downs in the southern counties. The surface of the chalk is dry, with little bare rock visible. The thin soil is generally covered with a springy turf, except where the land is farmed. The hills often form cuestas, having gentle *dip slopes*, following the dip of the strata, and abrupt *scarp slopes*, where the chalk come to an end. Chalk scarps are more rounded than limestone scarps, owing to the absence of joints.

Exercises

1 For *three* of the following, name an example you have studied and describe its formation:
- (*a*) Waterfall
- (*b*) Delta or alluvial fan
- (*c*) Meanders
- (*d*) Gorge

2 Study figure 7.6, which shows a river basin.
- (*a*) Which tributary is least likely to dry up during a long dry spell?
- (*b*) Which area is most at risk from flooding?
- (*c*) What would be the effect on flooding risk of the planting of trees on the slopes and on higher ground?

Figure 7.11 A section of chalk to show the water table

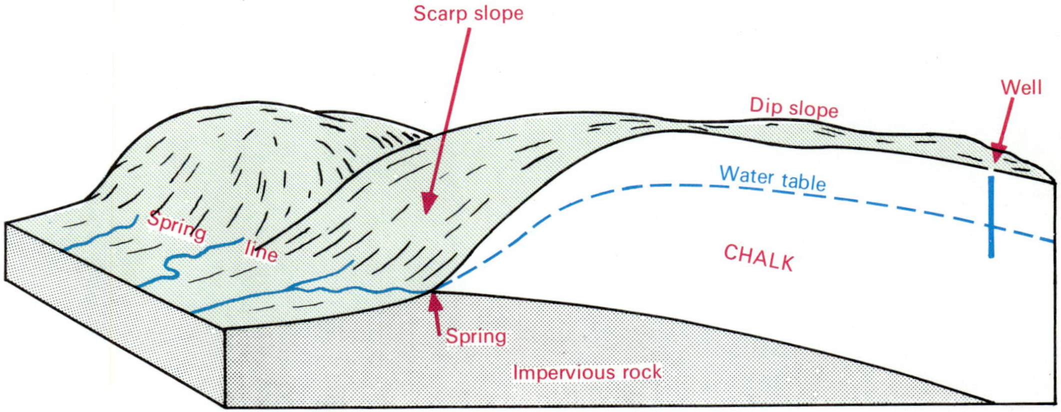

Chapter 8

Earth Sculpture in Arid Regions

A considerable part of the Earth's land surface receives so little rain that there are no permanent streams and very little natural vegetation. These regions are called deserts and the land forms that develop there are very different from those of more humid regions. They can be broadly classified into: (1) *rock deserts*, called *hammada* in the Sahara, (2) *stony deserts*, including both gravel and pebble types, and (3) *sandy deserts*, called *erg* in the Sahara.

Weathering

Because of the absence of cloud, the exposed rock is greatly heated by day but loses its heat rapidly by night. Expansion of the rock opens up cracks in the rock so that rainwater, however rarely it occurs, can find its way in. Evaporation leaves salt crystals behind, preventing complete contraction. Over time the process is repeated, these salt crystals grow and the rock splits and disintegrates.

The products of this disintegration range in size from large rock fragments to sand and fine dust. The coarser fragments do not usually travel far, but the sand is blown away by the wind, either along the ground, or in a series of hops. Dust travels in the air, sometimes over enormous distances. The most striking example of this is the *loess* of north-west China. This is a fine dust, hundreds of metres thick, which has been carried by wind over the centuries from the great deserts of central Asia.

The wind, when armed with sand grains, has considerable powers of erosion. This form of erosion is known as *abrasion* and has the greatest effect near the base of exposed rock, since most sand grains are near the ground. For this reason high ground exposed to wind abrasion has steep or vertical edges, often with a fretted surface due to variations in the resistance of the different strata, whilst isolated rocks may assume a mushroom or anvil shape.

Figure 8.1 Examples of weathering in arid areas

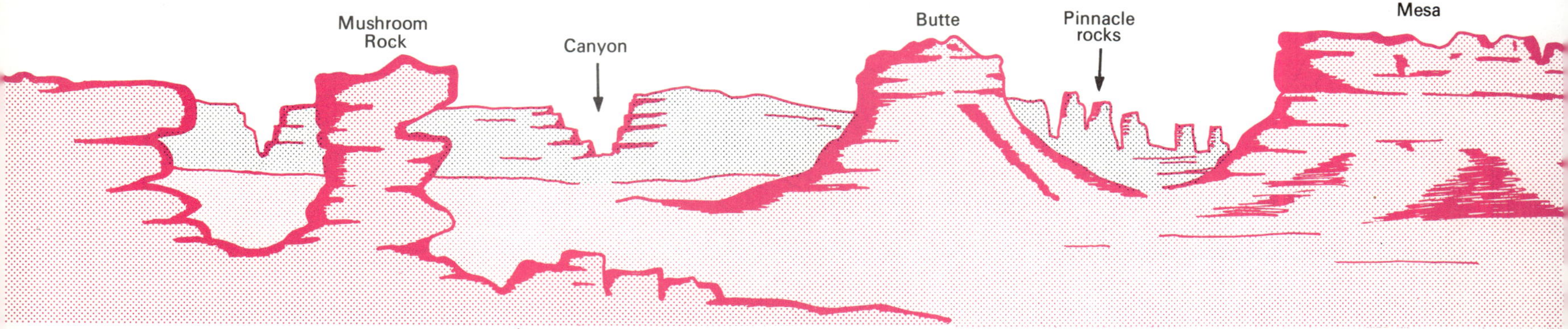

Sand Dunes

The sand carried away by the wind is later deposited, and in this way the great sand deserts have been built up. The sand is frequently heaped up into dunes, sometimes in a haphazard manner but often in a definite pattern. One common form is the crescent-shaped *barchan* which develops where the wind blows steadily in one direction. As the dune slowly advances the sand is blown up it but forms a steeper 'slip-face' on the lee side. The sides, offering less resistance to the wind than the middle part of the dune, move forward more rapidly; hence the crescent shape. Groups of barchans may cover many square km. Another type of dune is found where the prevailing wind is occasionally interrupted by cross-winds. The latter prevent barchans from forming and the sand is blown into long, narrow *seif* dunes, parallel to the prevailing wind.

Surface Water

Few deserts are completely rainless and the occasional showers can have a great effect on a desert surface which is unprotected by vegetation. Hillsides become deeply dissected by gullies and ravines, and in areas of less resistant rock a form of '*badland*' scenery develops, in which the gullies are so numerous as to render the land almost inaccessible. A striking feature of many deserts is the dried-up river valleys, called *wadis* in the Sahara. Most of these have flat floors and their sides, largely unaffected by rain, are steep and sharp edged. When rain does fall they are rapidly filled but the water quickly seeps away and the river often ends up as a pasty mass of sand.

Large perennial rivers sometimes rise in humid regions and flow across deserts. An example is the River Nile, which rises on the East African Plateau and later flows across the Egyptian desert, where it is used extensively for irrigation. Another example is the River Colorado in south-western United States. Here the river has cut the Grand Canyon, a gorge $1\frac{1}{2}$ kilometres deep. The absence of rain has helped to reduce the effects of weathering on the valley sides which rise by a succession of steps and slopes depending on the varying resistances of the strata.

Rain falling on distant mountains may move through a layer of permeable rock and reach the surface in a desert, giving rise to an *oasis*. The water may gush out as an *artesian well*. These occur in the lower part of a syncline, the water rising because of the pressure exerted by water at a higher level, as shown in figure 8.3. Of course artesian wells occur in many regions other than desert lands.

Within many deserts there are broad *basins of inland drainage*, into which flow a number of intermittent streams. These basins frequently con-

Figure 8.2 Types of sand dune

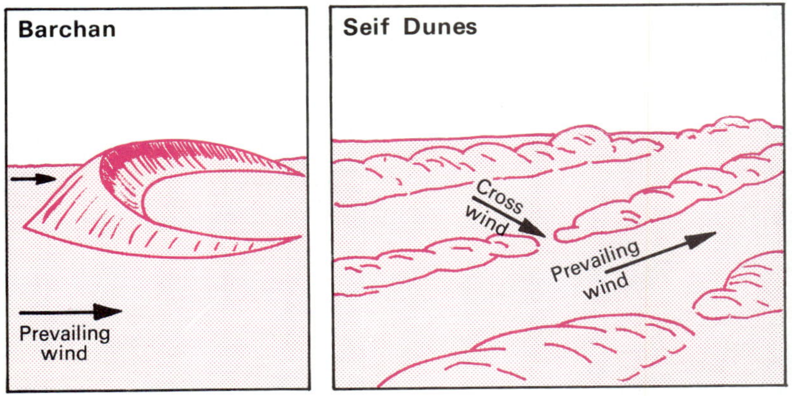

Figure 8.3 An artesian basin

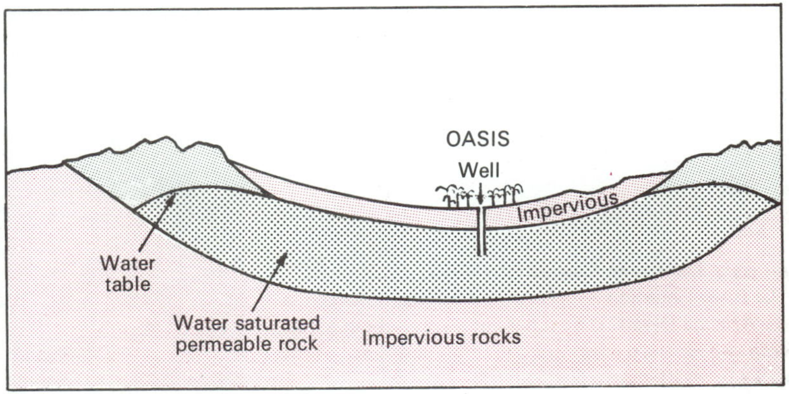

tain one or two small salt lakes, or a sheet of salt called a *salt pan*. The water escapes by evaporation, leaving behind any dissolved salts. Lake Eyre, in the Australian desert, is a large lake of this type.

Residual Masses

The various forms of erosion mentioned earlier in this chapter gradually wear back and dissect the high land until only the more resistant rocks remain as isolated hills or mountains. These residual masses usually consist of resistant beds overlying and protecting less resistant beds. Where the beds dip in one direction scarp and dip slopes are formed, but where they are horizontal a flat-topped hill, surrounded by escarpments on at least three sides, is found. In the North American deserts this type of flat-topped hill is called a *mesa* (Spanish for table), or where the summit has become relatively narrow, a *butte* (figure 8.1).

Dunes in Death Valley, California ▶
(*Aerofilms*)

far right **The Grand Canyon** (*Aerofilms*)
The River Colorado is a perennial river and flows across the Arizona Desert. Here it has cut a gorge about a mile deep, whose steep edges are due to the resistance of the rocks to weathering.

SOIL EROSION

The loss of soil, due to erosion, is a serious problem in both arid and humid regions, but it is particularly so in areas of light rainfall, where long droughts are common and the rain falls in heavy showers.

The soil is probably our most valuable asset, since it supports both plant and animal life. It has taken thousands of years to form and consists mainly of very fine rock particles, combined with organic matter usually in the form of humus. A fertile soil contains such essential plant foods as nitrates, potash and phosphates, and also is able to hold water for long periods.

Unfortunately, people have been misusing the land and causing a considerable loss of valuable soil. This has been brought about by deliberate deforestation and by allowing large numbers of goats and other animals to graze in woodlands. The animals devour all the young shoots so that when the old trees die there are none to replace them. The natural vegetation is nature's own protection for the soil and if it is removed the soil is exposed to the ravages of wind and rain. Where there are steep slopes the rain quickly cuts gullies, which are rapidly enlarged. If the land has been ploughed directly up and downhill the rainwater rushes down the furrows, carrying soil with it. There is also a much slower, almost imperceptible downhill movement of soil particles, known as *soil creep*. Even on flat land the growing of exhaustive crops like cotton or maize, year after year, without the regular application of manure, reduces the humus content of the soil. This makes the soil dry and dusty, and liable to be blown away by the wind.

A large area in the Middle West of the United States, between the Mississippi valley and the Rockies, became a 'dust bowl' in this way, and many farmers had to leave their homes. Another badly eroded area was the Tennessee valley, also in the United States. Here the population was reduced to poverty, partly as a result of soil erosion. The United States Government set up the Tennessee Valley Authority, with instructions to restore the fertility of the soil and to help the population to make the best use of the land. The necessary measures included the large-scale planting of trees, the filling in of gullies and the construction of numerous dams on streams and rivers, so as to control the flow of water (the dams have also made water available for hydro-electricity). Help and advice were given to the farmers, to persuade them to make a greater use of fertilisers and manures, and to adopt a system of crop rotation aimed at keeping the soil in good condition. Where sloping land was under the plough, *contour ploughing* was recommended, which means ploughing along the contours instead of up and down the slopes so that the danger of gullying during rainstorms is reduced. The success of the Authority's work has made the Tennessee valley a model for similar schemes elsewhere.

Desertification

In recent years a new word has been coined to describe the way some areas of the world are turning into deserts—desertification. The parts of the world that are most at risk are the dry lands which have long dry seasons and fluctuating rainfall. About 30% of the world's land surface is affected and one-sixth of the world's population (over 650 million people).

The problems already mentioned in this chapter apply to the process of desertification, but there are other matters to consider. Loss of vegetation cover is often caused by the need to collect firewood for cooking, or by the shortening of a fallow period because of population pressures, or by trampling when herds follow regular routes to waterholes. Irrigation brings about problems of waterlogging and salinisation (the accumulation of salt in the soil). Shortening of a fallow period also leads to soil exhaustion and the breakdown of soil structure, so the soil blows away as described earlier.

There are many individual methods of dealing with the effects of desertification. Cost is an important factor, since the worst-affected areas include the poorest countries of the world. The most important methods include (*a*) controlling the use and flow of water to reduce waterlogging and salinisation, and (*b*) conserving soil by keeping a vegetation cover, fencing off areas from grazing animals and terracing slopes.

Gullying (*D Jones*)
Suggest how the gullying could have been prevented and how its effects could be remedied.

Dust Bowl (*U S Information Service*)

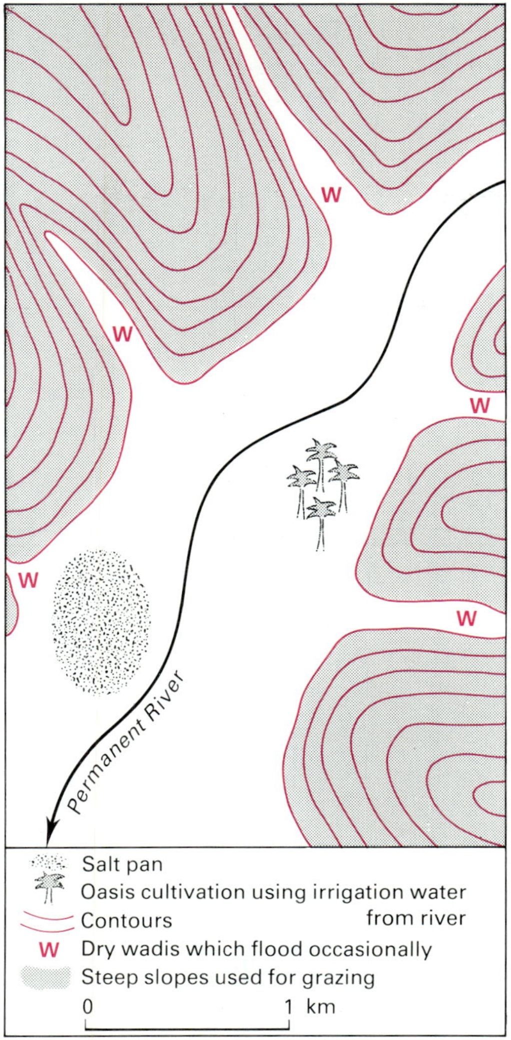

Figure 8.4 A semi-arid area

Exercises

1 Only about one third of desert areas are covered by sand.

(*a*) Explain how the sand itself is formed.

(*b*) With the aid of sketch diagrams, describe and explain the formation of sand dunes.

(*c*) Describe the action of wind as an agent of erosion.

2 (*a*) What is a *wadi*?

(*b*) Even though rainfall is rare there is rapid erosion by water when it does rain in a desert. Explain why this is so.

3 Study figure 8.4 which shows a semi-arid area where desertification is a major risk.

(*a*) Why are the slopes at risk? Suggest action to reduce the risk.

(*b*) What could be done to reduce or eliminate the damage done by occasional floods in the dry wadis?

(*c*) If irrigation was extended on the flat land, what might result from excessive evaporation of irrigation water?

Chapter 9

Glaciers and Ice Sheets

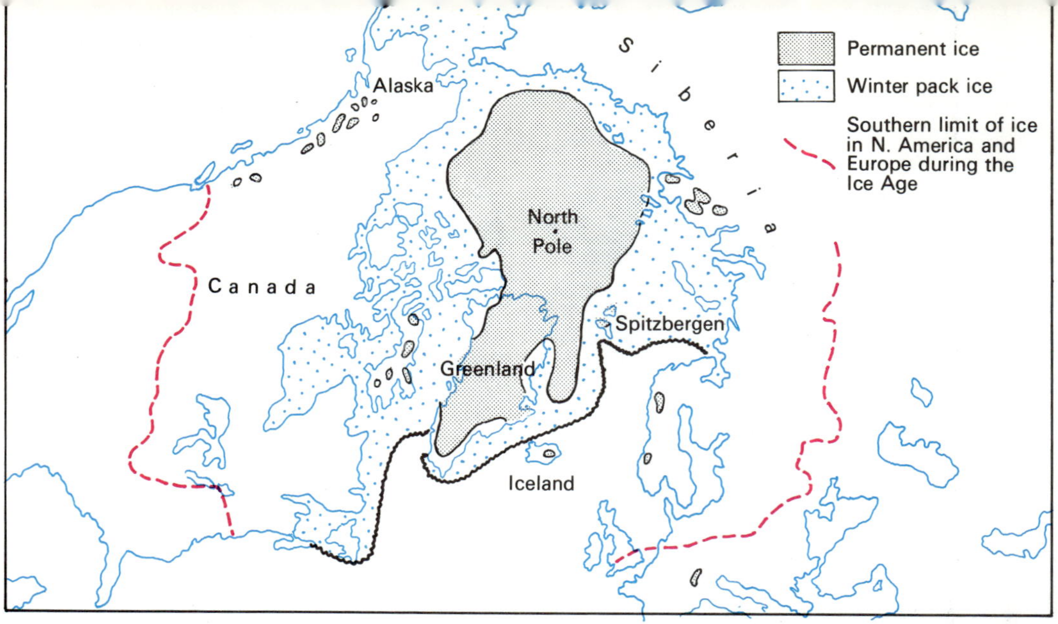

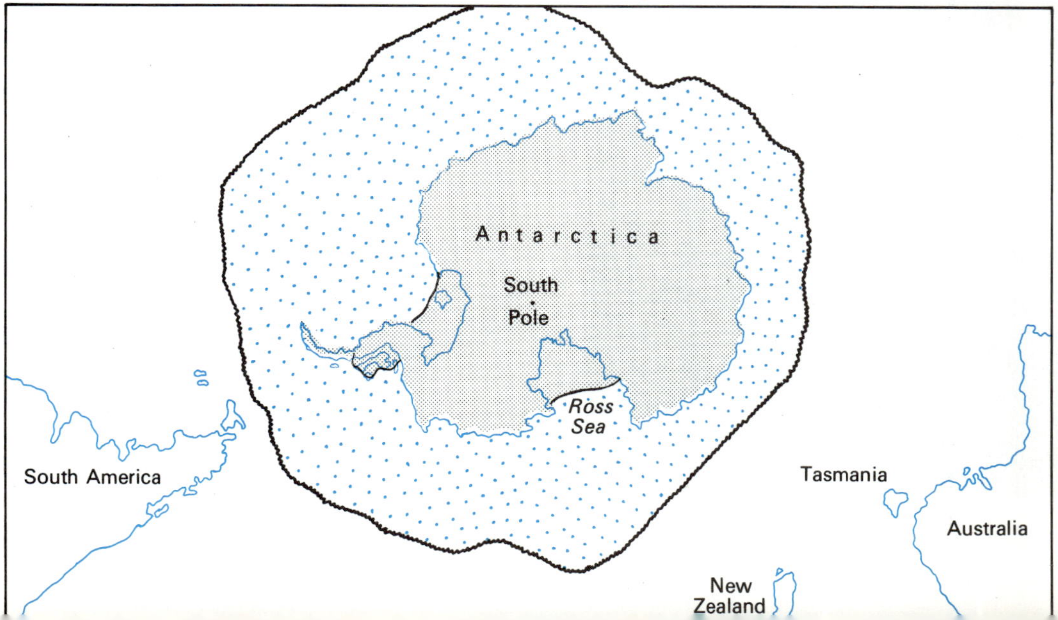

Where snow lies to a great depth its lower layers are compressed into a compact but granular form of ice, whilst at greater depths still solid ice is produced. The ice is capable of movement, normally in a downhill direction, especially when under pressure from ice lying at a higher level.

Large areas of the world are permanently covered with snow and ice, almost all in the Arctic and Antarctic regions, where the precipitation takes the form of snow and the low temperatures prevent any appreciable thawing. In Antarctica the land surface is largely under an *ice cap*, though high mountains pierce the ice in some places. The ice cap feeds numerous glaciers and ice sheets which flow down to the sea. In the Ross Sea an immense floating ice sheet called the Ross Barrier ends in an ice wall between 15 and 60 metres in height. Much of Greenland is also covered by an ice cap, whilst there are smaller ones in Iceland and Spitzbergen. Where glaciers and ice sheets reach the sea *icebergs* are continually breaking away. These are a serious menace to shipping with only one-ninth of the berg showing above the water, though the danger is less today than it used to be as a result of the use of radar.

In other parts of the world permanent accumulations of snow and ice are found only at high altitudes above the *snow line*, which increases in height from sea level at the Poles to about 1000 metres in northern Norway, 3000 metres in the Alps and nearly 6000 metres on the Equator. No part of the British Isles reaches the snow line, although patches of snow may be seen even in summer on the northern slopes of Ben Nevis and other high mountains.

◀ **Figure 9.1 The earth's permanent ice cover and winter pack ice**

Cirques and Glaciers

Most great mountain ranges carry extensive snowfields on their upper slopes. The snow slides off the steepest slopes but accumulates on the gentler slopes, and particularly in the hollows, where it becomes compressed into ice. These hollows are enlarged by the action of the ice to form *cirques* (which are also known as *corries* or *cwms*), with steep rock walls on three sides and a lower lip on the fourth, over which the ice escapes. The enlarging process is as follows: during the daytime, especially in summer, meltwater penetrates beneath the ice, usually down a gap called the *bergschrund*, and soaks into the underlying rock. By night this water freezes, causing the rock to break up. The loosened fragments are eventually carried away by the ice as it slowly creeps forward, under pressure from fresh snow.

Figure 9.2 Section of a cirque

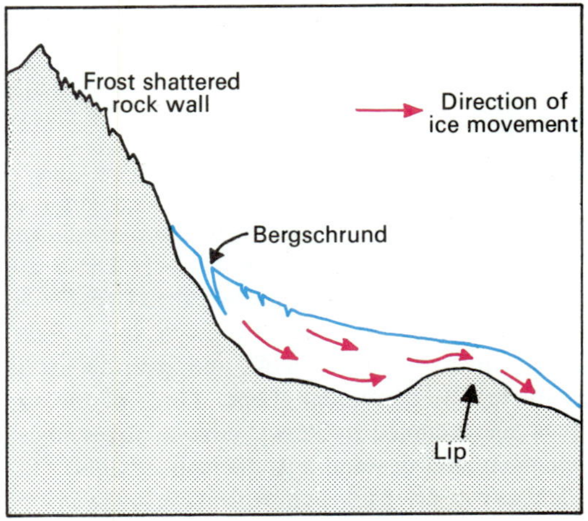

Cirques often surround a mountain summit, and as they grow in size the watersheds between them are narrowed down until they form sharp-edged ridges, known as *arêtes*. A number of arêtes often converge on a summit, giving the latter a pyramidal shape. Many Alpine peaks are of this type, the Matterhorn being the most striking example.

From the cirques the ice moves downhill to join a *valley glacier*. A glacier is a vast river of ice, hundreds of metres thick, which moves only a few metres per day, and even slower near the edges where it is retarded by contact with the valley sides. The surface of the ice is often uneven, with deep *crevasses* caused by variations in the gradient or width of the valley. A steep gradient may produce an *ice fall*, with the ice breaking up into great blocks which reunite below the fall.

The glacier, armed with fragments of rock that it has picked up, produces an abrasive action on the valley floor and sides. This abrasion is combined with a plucking action, by which loose fragments become embedded in the ice and are removed. In this way the valley is deepened and widened, leaving it with steep sides and a flat base. Such a valley is termed as *U-shaped valley*.

The particles of rock picked up by the ice are carried with it and form *moraines*. The material which falls from the valley sides on to the ice forms *lateral moraines*. Where glaciers join, the inside lateral moraines unite and form *medial moraines*, whilst at the base of the glacier large quantities of fine-grained 'rock flour', together with larger fragments, constitute the *ground moraine*. As lower altitudes are reached an increasing amount of melting occurs and morainic material may cover much of the ice. Where the glacier ends the larger rocks are deposited and form a crescent-shaped *terminal moraine*. The finer gravels and sands are carried some distance by the meltwater streams before they are laid down as part of the *outwash plain*. These streams usually flow out of the ice at a cave or *snout* and are discoloured by fine clays carried in suspension.

All these features are represented in any of the larger Alpine glaciers, the largest of which is the Aletsch glacier, which is about 25 kilometres long and $1\frac{1}{2}$ kilometres wide. The thickness of ice in it reaches 800 metres in places. Other well-known glaciers are the Mer de Glace and Rhône glacier.

THE QUATERNARY ICE AGE

About 800 000 years ago, in the Pleistocene period, temperatures began to fall and ice slowly advanced from the north until it covered considerable areas in northern Europe and North America. In Europe it spread mainly from Scandinavia and the Alps over most of the north European plain. From Scandinavia it crossed the North Sea to the British Isles, which had its own centres of glaciation in Scotland, the Lake District, Northern Ireland and Wales. The southern limits of the ice in Britain followed roughly the present valleys of the Thames and lower Severn. Southern England remained unglaciated and there is evidence here that early people inhabited caves which provided some protection from the cold.

The Ice Age was interrupted on at least three occasions by *interglacial periods* when mild conditions returned. It has been estimated that the present mild period began about 20 000 years ago. With temperatures in northern Europe still rising slightly, and both the Alpine glaciers and Polar ice caps in gradual retreat, it seems unlikely that cold conditions will return in the near future.

During the glacial periods the great weight of the ice pressed down the surface of the land, but in the inter-glacial periods much of the ice melted, raising the level of the sea, and this was followed by a slow uplift of the land, relieved of its burden of ice. The drowned coastlines and raised beaches of northern Europe and North America are visible evidence of recent movements of this type.

EVIDENCE OF PAST GLACIATION IN THE BRITISH ISLES

In Mountain Areas

One of the most typical glaciated mountain regions is Snowdonia, in North Wales. The summit of Snowdon is flanked by four steep-sided cirques (or cwms, as they are called in Wales). These are separated by long, jagged arêtes, called Crib Goch, Lliwedd, Bwlch Main and Clogwyn du'r Arddu, which meet at the summit and give to it its characteristic pyramidal shape. Three of the cirques are occupied by small, circular lakes, from which streams descend into the valleys below. One such lake is called Glaslyn. It lies just to the east of Snowdon and has a maximum depth of 38 metres. From it a stream descends into another lake, Llyn Llydaw, 58 metres deep, whose waters are partly held up by a morainic dam. This moraine undoubtedly represents a late stage in the retreat of the glacier, towards the end of the Ice Age.

Other striking features are the deep, glaciated, U-shaped valleys which surround Snowdon on three sides. These are the Llanberis Pass in the north, the Nant Gwynant in the south-east and Nant Betws in the west. Glaciers have straightened the sides of these valleys, shearing away any projecting spurs and the lower parts of tributary valleys. The worn-back remnants of such spurs are called *truncated spurs*, and the valleys that now enter the main valleys at a high level are called *hanging valleys*. The streams which occupy these hanging valleys often form torrents or waterfalls.

The floor and sides of the main valleys show evidence of abrasion by ice, in the form of smoothed rock surfaces, including *roches moutonnées* (rocks shaped like the back of a sheep) with a smooth slope facing up the valley and a rough craggy surface facing downstream. Some of these rocks bear scratches (*striations*) made by rock fragments embedded in the ice.

These glaciated valleys also contain a number of long, narrow ribbon lakes, some of which occupy rock basins scooped out by the glaciers whilst others are held back by morainic dams. In the Llanberis Pass are two lakes of the first type, Llyn Peris and Llyn Padarn. These were originally one lake but have been separated by an alluvial plain deposited by the River Arddu.

Similar evidence of glaciation can be found in other mountain areas. In the Lake District, the slopes of Scafell, with the adjoining valley of Borrowdale, and the eastern slopes of Helvellyn, including the remarkable arête of Striding Edge, are fine examples. There are many glaciated areas in Scotland, particularly in the Cairngorms. The existence of many drowned glaciated valleys (fiords) along the coast of north-west Scotland has already been mentioned in Chapter 7.

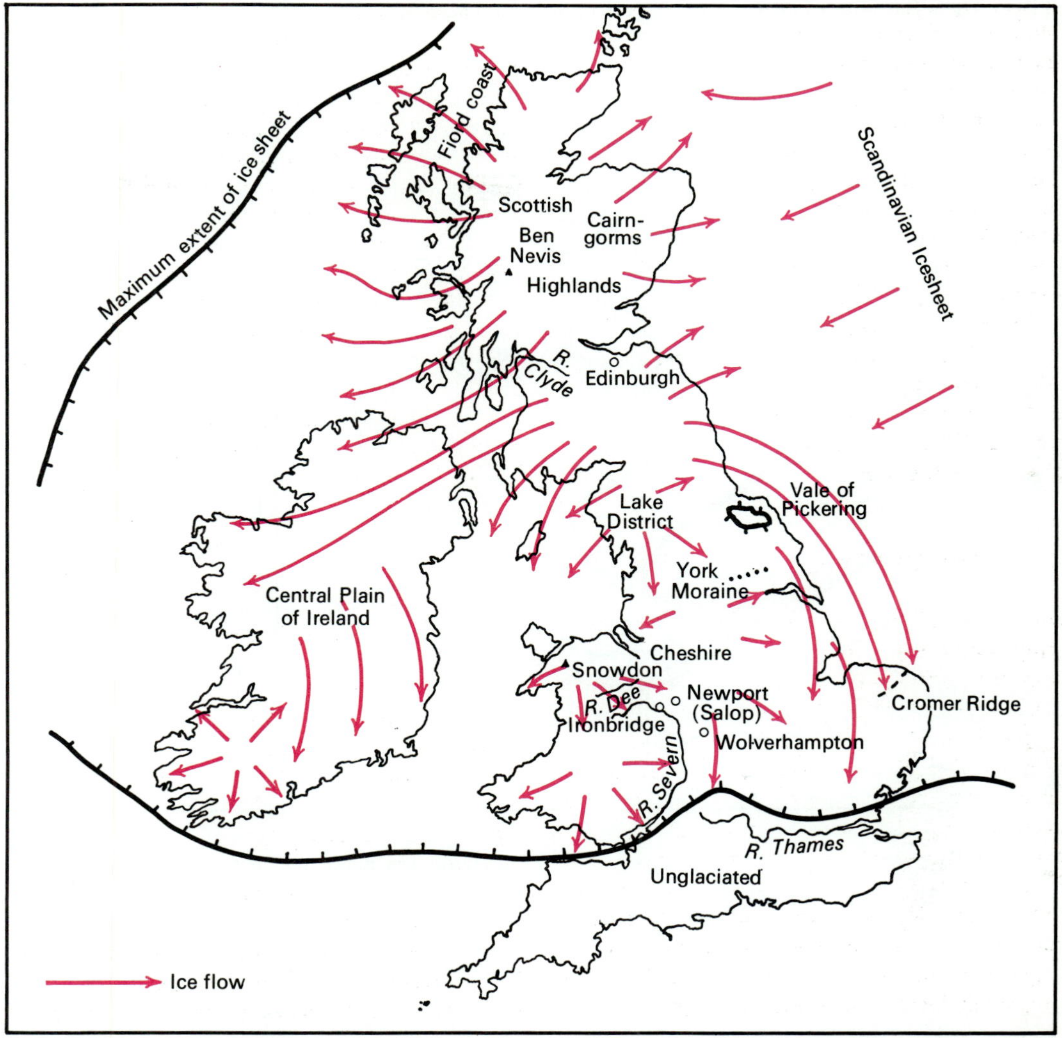

Figure 9.3 The British Isles in the Ice Age

Figure 9.4 A glacier in an alpine valley and the valley after glaciation

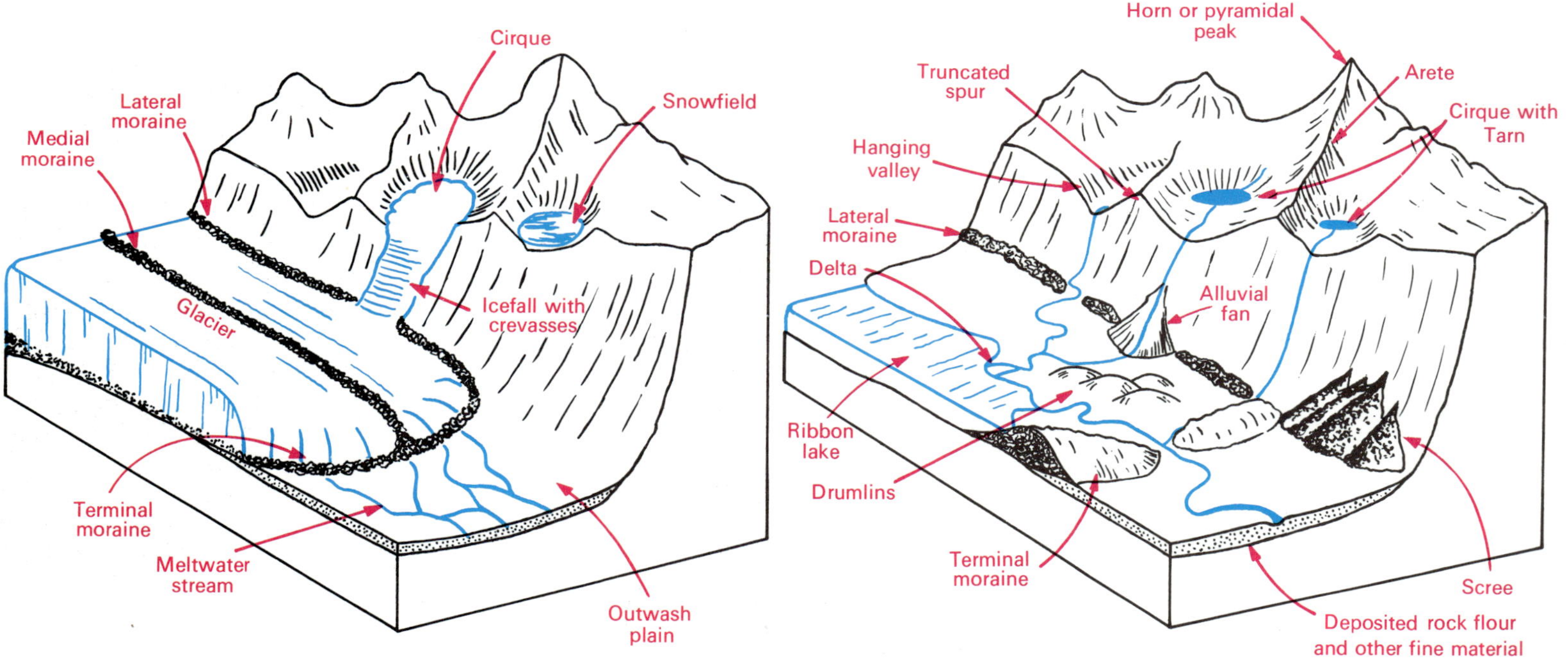

facing (*a*) **The upper half of a glacier,** (*b*) (*right*) **the lower half of a glacier** (*E G Bowley ARPS*)
(*a*) The Alpine glacier is 'fed' from many cirques. Note the snowfields, arêtes, pyramidal peak and the lateral and medial moraines. (*b*) The glacier shrinks as it approaches its melting point, and the moraines spread out over much of the ice. Meltwater from the snout has formed a lake.

Llanberis Pass (*F R Dobson*)
Make a line drawing of the scene, labelling the screes, truncated spur, stream and large roche moutonnée.

Why is the valley U-shaped, and how was the roche moutonnée formed? What has happened since the Ice Age to modify the valley shape?

a

b

In Lowland Areas

Most of the lowlands of the British Isles, except for England south of the Chilterns and Bristol Channel, were covered with ice sheets during the Ice Age. Much has been learnt of the movements of these ice sheets from the distribution of *erratics*. These are rocks which have been transported by ice from their place of origin and deposited elsewhere; for example numerous erratics from Ailsa Craig, in the Firth of Clyde, are found scattered widely over Wales and Ireland.

Many low-lying glaciated areas are covered by deposits of *boulder clay*, a stiff clay which contains numerous stones and boulders. This was originally the ground moraine of the ice sheets and sometimes occurs in rounded hummocks called *drumlins*. Extensive deposits of sand and gravel were also left behind, mostly from the moraines and outwash plains of the ice sheets. The Cromer Ridge in Norfolk was the terminal moraine of the North Sea ice sheet, whilst another terminal moraine forms a ridge about 15 metres high running from east to west across the Vale of York and passing through York itself. Other smaller ridges of sand or gravel known as *eskers* occur in central Ireland and between Wolverhampton and Newport (Shropshire). These are thought to have once been the beds of streams that flowed at the base of the ice, the sand and gravel being left behind when the ice melted.

Silvaplana, Switzerland (*Aerofilms*)
Draw a sketch map of this area to show the tributary, U-shaped valley, hanging valley, delta and the part of the lake in the foreground. Explain the origins of the hanging valley and delta.

The boulder clay forms an impervious layer so that the surface has many streams, lakes and marshes. Lakes have been caused by the clay blocking former valleys or merely by water occupying hollows in it, as in some of the Cheshire meres.

Figure 9.5 The Vale of Pickering

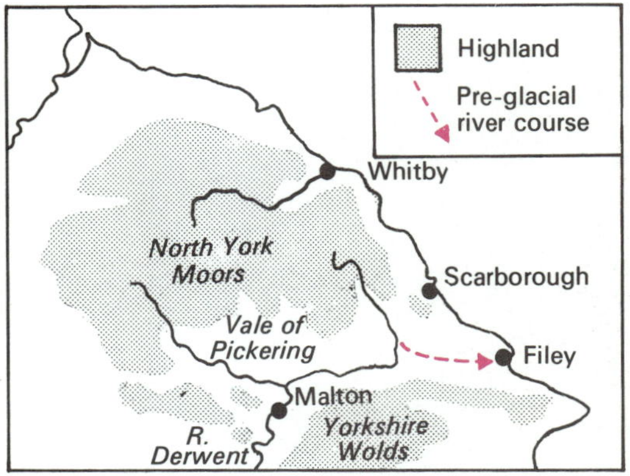

◀ **A Norwegian fiord, a drowned glaciated valley** (*Aerofilms*)

During the Ice Age lakes accumulated against the margins of the ice sheets and developed *overflow channels*, some of which are occupied today by rivers. In the Vale of Pickering, North Yorkshire, the North Sea ice sheet blocked the original outlet of the River Derwent at Filey, causing a large lake to form between the North York Moors and the Yorkshire Wolds. An overflow channel formed in the south-west which in time cut the present Malton gorge. After the ice retreated the River Derwent retained its new course. Similarly the upper Severn, which before the Ice Age flowed into the present Dee, was diverted by ice and has cut the Ironbridge gorge, through which it now flows southwards.

Another feature of glaciated lowlands is the low, rounded hills which have been over-ridden and smoothed by ice. In contrast some higher hills, like the volcanic mass of Castle Rock on which Edinburgh Castle stands, have a *crag and tail* formation. The ice, failing to surmount Castle Rock, divided and went round it, producing an abrupt crag facing the direction of the ice movement and a gently sloping tail on the lee side.

Exercises

Study photograph **b** on page 93.

1 (*a*) Make a simple sketch of the scene. Add labelling to show: glacier snout, melt-water lake, moraine, zone of accumulation, crevasses.

(*b*) What evidence is there to show that the glacier used to be larger?

2 Compare and contrast a typical river with a typical glacier, under the following headings: flow, valley shape, erosion and deposition.

Chapter 10

The Work of the Sea

The edge of the land, where it meets the sea, is known as the *coast*. Along the coast there may be mountains, hills or plains. The higher land usually consists of hard rock which forms cliffs and promontories, whilst the lower land is frequently of softer rock and slopes gently to the sea. The sea itself is constantly at work, buffeting and wearing back the land in some places, and adding to the land by depositing material elsewhere.

Wave Action and Beach Formation

The sea is seldom still. Its level rises and falls with the tides, which also produce tidal currents along our coasts. But *waves*, caused by the wind, exert the most powerful influence on the coast. Out at sea each particle of water in a wave follows a roughly circular path, returning to its former position. This is why a piece of wood moves forward with the crest of a wave then backward with the trough, and so does not travel onward with the wave. When a wave approaches a shoreline its lower part is slowed by friction with the sea bed. The upper part continues to move forward and, being unsupported from below, develops into a *breaker* which, as it topples over, causes a foaming mass of water to surge up the beach.

Beaches are found between and around the high- and low-water marks and consist of debris derived from the land. The advance of water up a beach, following the breaking of a wave, is called the *swash*, and the return movement is the *backwash*. The swash pushes some of the beach material landwards, whilst the backwash carries it back to the sea. But waves vary greatly in strength and direction. Where there is a heavy ground swell, probably the result of some distant storm, the

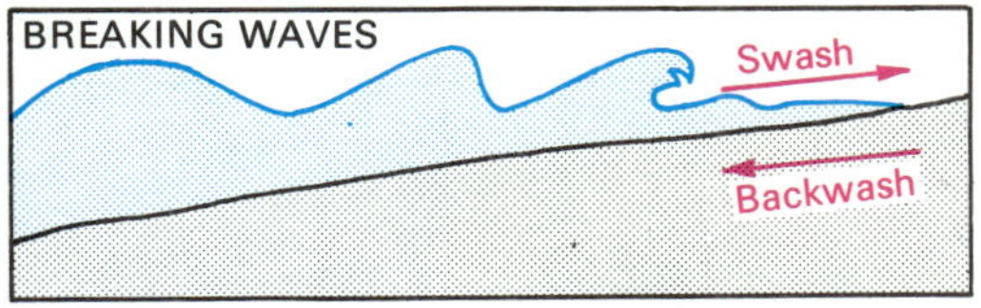

Figure 10.1 Breaking waves

◀ **Cliffs under wave attack** (*E G Bowley ARPS*)

waves are long in proportion to their height, and are in the main *constructive* in their effects. On the other hand severe local storms produce waves that are high in proportion to their length, so that as they break they plunge almost vertically and are highly *destructive*. This type of wave can measure up to 15 metres from crest to trough and on breaking may produce a pressure equal to about 20 tonnes per square metre on exposed rock.

The beach material is also graded by wave action, with the larger and more angular blocks of rock nearest the land. Then come smaller, rounded stones called *shingle*, formed by the break up of rocks and the knocking together of fragments as a result of wave action. Lower down the beach sand is often found, and sometimes silt. A current flowing along the sea bed away from the shore, known as the *undertow*, is responsible for moving much of this finer material out to sea.

Figure 10.2 Beach deposits

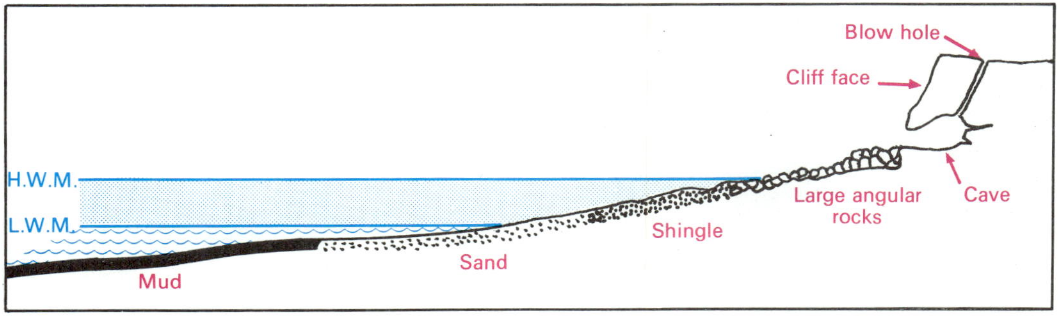

Cliff Coastlines

Where high land forms the coastline, the action of the sea on the land leads to the formation of cliffs. The pounding of the waves, and the cutting action of rock fragments which they carry, undermine the cliffs so that the upper part, which is subject only to normal weathering, falls to the beach. As the cliffs recede a wave-cut platform is left behind, partly covered by beach deposits. In time this may become so wide that it checks the speed of the advancing waves and the cutting back of the cliffs may practically cease.

Cliffs composed of relatively soft rocks, like the sand, gravel and boulder clay cliffs of Holderness and East Anglia, offer little resistance to wave action and are subject to frequent landslips. In Holderness the coast has been worn back nearly 5 kilometres since Roman times, with the loss of some thirty small towns and villages.

In contrast, cliffs of hard rock like limestone or granite recede slowly. Fallen blocks often protect the cliff base from the full fury of the waves. But in well-jointed rocks water is forced into the crevices, compressing the air trapped within them, so that they are enlarged. In this way *caves* are formed and sometimes, as in parts of Cornwall, they have been extended until *blow holes* open on the cliff top. Where headlands are eroded from two sides, a *natural arch* may be produced and with continued erosion the arch may collapse, leaving a *stack*. The Needles, at the western end of the Isle of Wight, and the Old Harry Rocks near Swanage were formed in this way.

If the beds dip seawards landslips are liable to occur. These are especially frequent along the south coast of the Isle of Wight and between Dover and Folkstone in Kent, where the chalk overlies a bed of slippery clay.

Differential Erosion

Where rocks of varying resistance occur along a coast the sea generally wears back the softer rocks to form bays, whilst the harder rocks stand out as promontories. After a time the deposition of shingle and sand in the bays protects them from wave action and they are no longer enlarged. Some fine examples of this differential erosion are found in Dorset. North and south of Swanage the coastline is of a *transverse type*, which means that the ridges run roughly at right angles to the coast. A northern ridge of chalk and a southern one of limestone terminate in rocky headlands, whilst a clay vale separating them forms Swanage Bay.

Figure 10.3 Features along the south coast of England

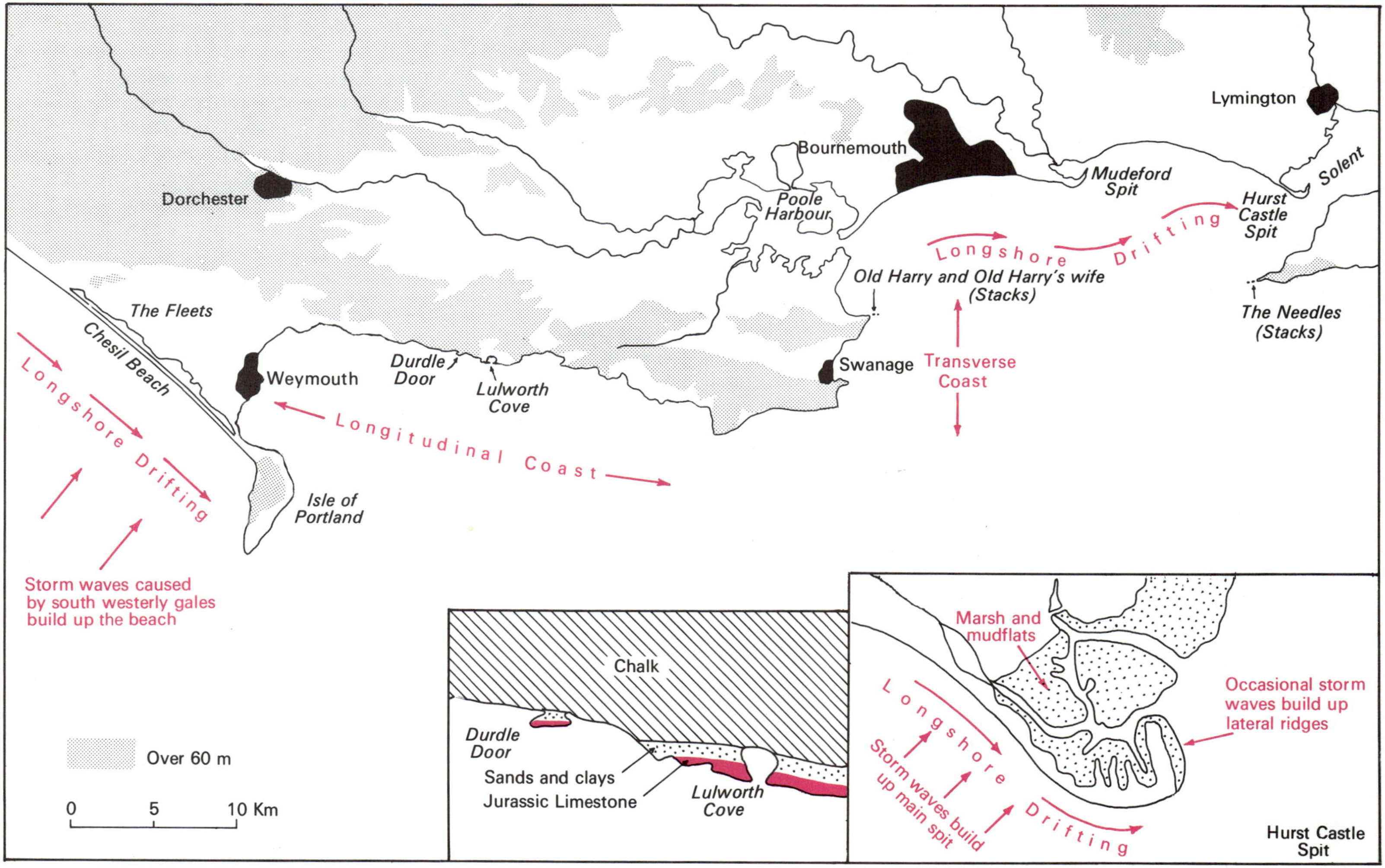

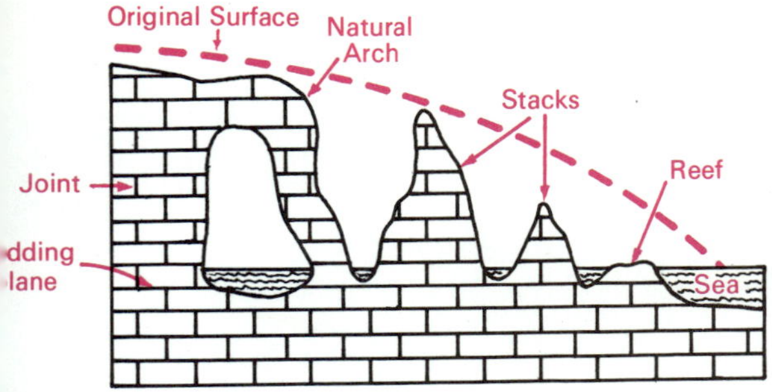

Figure 10.4 Recession of a promontory

Farther to the west around Lulworth the coastline is of a *longitudinal type*, with the ridges parallel to the coast. The coastal limestone ridge forms high cliffs. At Lulworth Cove the sea has broken through the limestone and eroded a circular bay in the softer clay, but is prevented from working farther inland by massive cliffs of chalk. A little more to the west a similar process is beginning at the Stair Hole, and farther west still the limestone

Figure 10.5 Longshore drifting

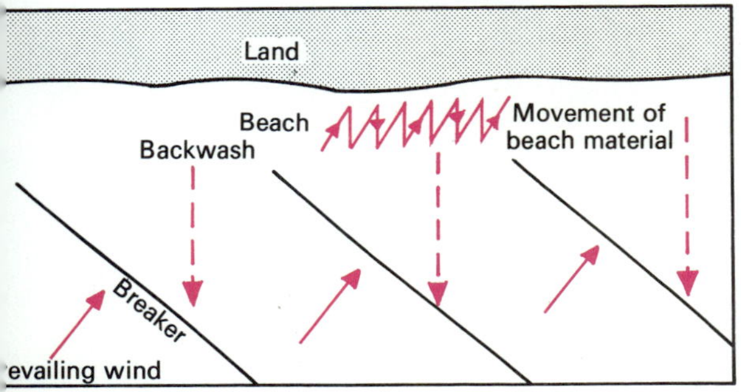

passes out to sea, where, before it finally vanishes, it forms a magnificent natural arch called Durdle Door.

The Longshore Drift

If waves are driven by the prevailing wind to approach a shore obliquely, sand and pebbles are washed obliquely up the beach, but return with the backwash straight down the beach. Each pebble thus follows a zigzag course and the total effect is a gradual movement or *longshore drift* of material along the coast.

Much of the south coast of England is affected by a west to east longshore drift, since the majority of waves approach the land from the south-west. *Groynes* have been built at many holiday resorts to prevent the loss of beach material.

Spits and Bars

Where there is an inlet or promontory the sand or shingle moved by the longshore drift is often deposited to form a long narrow *spit* or, if it stretches right across an inlet, a *bar*. Powerful waves assist in building up spits and bars by throwing up the shingle well above the normal level of the high tide.

A fine example of a shingle spit is Hurst Castle spit, stretching into the Solent. This includes a number of lateral spits, almost at right angles to the main spit, caused by occasional north-east storm winds driving waves against the end of the main spit.

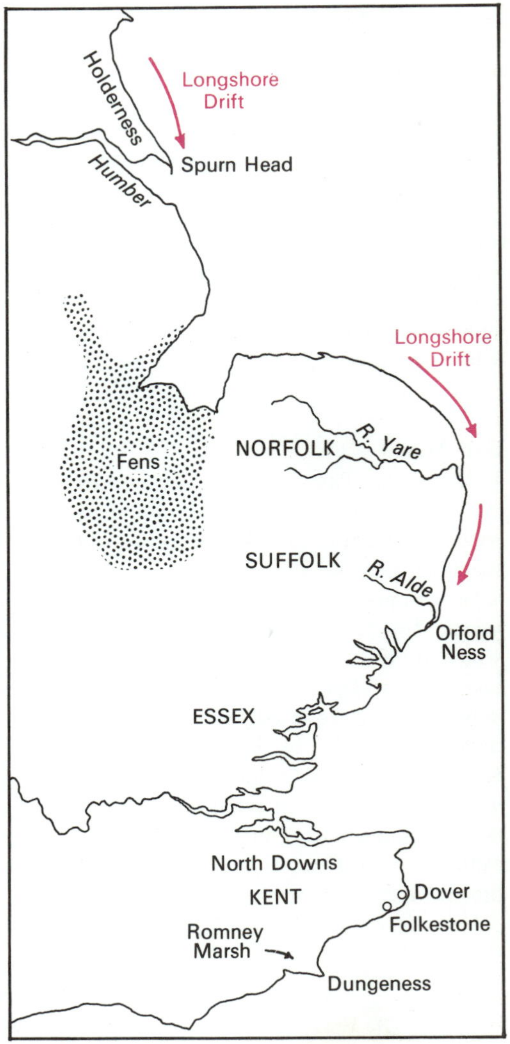

Figure 10.6 Features along the east coast of England

In Dorset there is the shingle bar of Chesil Beach, stretching for about 28 kilometres along the coast and linking the isle of Portland with the mainland. On the landward side of the bar is a long, narrow lagoon called the Fleets, occupying the hollow between the bar and the original coastline. Along the east coast of England, north to south tidal currents and a longshore drift have combined to form a series of spits, mostly across the mouths of rivers. These include Spurn Head, at the mouth of the Humber, another across the mouth of the Yare, in Norfolk, and the longest, Orford Ness in Suffolk, which has diverted the mouth of the River Alde southwards for a distance of about 20 kilometres.

Salt Marsh

In some coastal areas land is being gained from the sea. This is mainly due to the accumulation of silt in sheltered estuaries and behind spits and bars. On reaching sea level the silt forms mudbanks, and with the appearance of vegetation it is called *salt marsh*. The banks are often intersected by creeks which are filled with water at each high tide. At a later stage reclamation is sometimes possible, the new land being protected from flooding by embankments and drained by ditches. Romney Marsh in Kent was reclaimed in this way from extensive salt marshes which formed behind the series of shingle beaches known as Dungeness. An even larger example is the Fens, where reclamation has continued since Roman times.

Old Harry Rocks, near Swanage (*Aerofilms*)

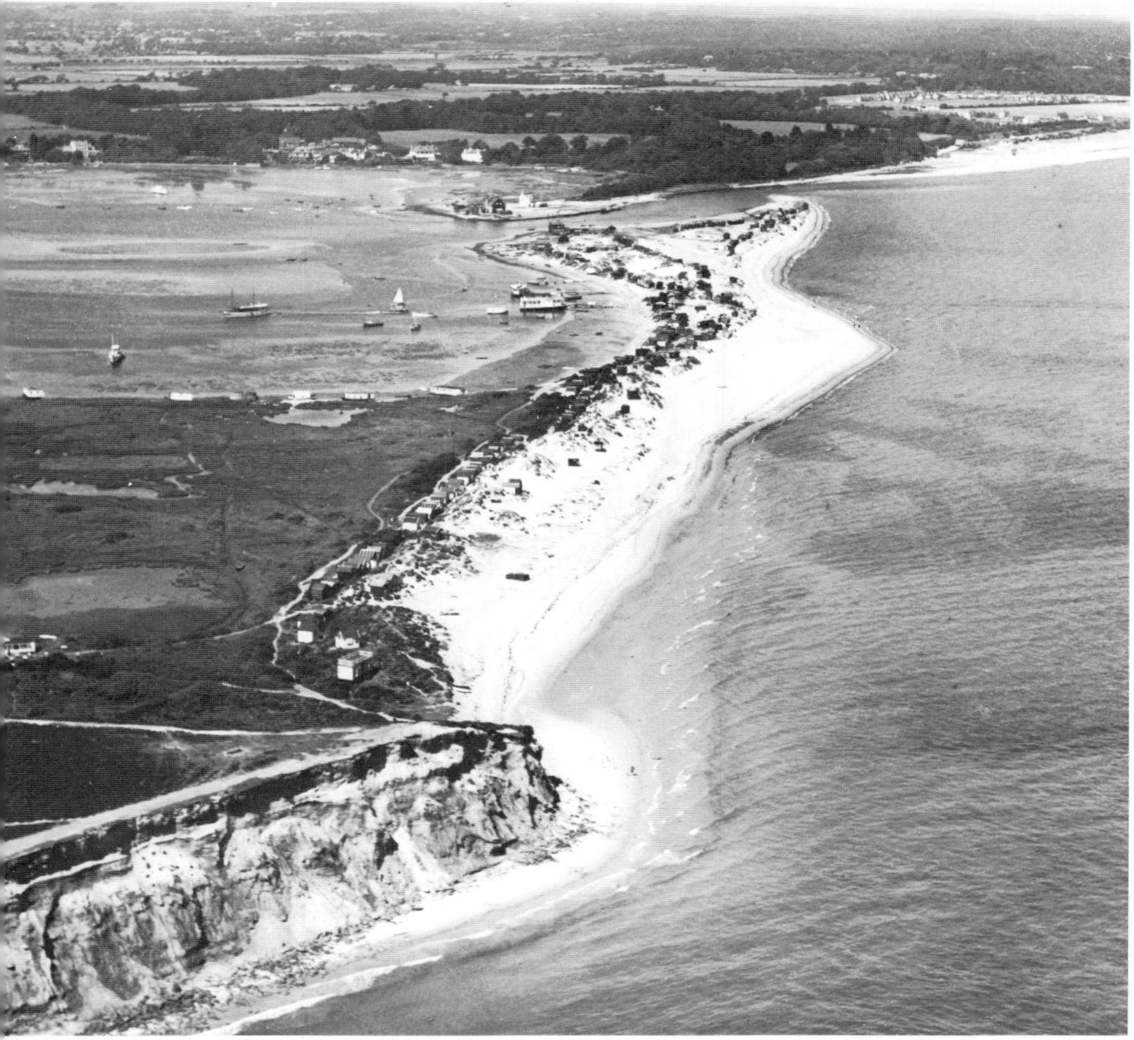

Mudeford Spit and Christchurch Harbour (*Aerofilms*)
(*a*) Make a line drawing of the Old Harry Rocks scene, labelling a stack and natural arch. Describe the stages in the recession of such a promontory. (*b*) Draw a sketch map of the Mudeford Spit area, naming Christchurch Harbour, the salt marsh and mudbanks uncovered at low tide. Explain how the spit was formed, and why Christchurch Harbour is silting up.

Coastal Sand Dunes

A prevailing onshore wind and a sandy beach are essential for the formation of *sand dunes*. The grains of sand are blown landwards and settle against any obstacles, including grass and trees. Sand dunes may advance inland, ruining crops and rendering the land useless. Special deep-rooted grasses, including *bent* and *marram* grass, help to bind the sand and prevent drifting, but it is often necessary to erect brushwood fences and plant belts of coniferous trees. This has been done near Formby and Southport on the Lancashire coast.

The Effect of Elevation and Subsidence on Coastlines

Where an elevation of the land has taken place the area of shallow water off the coast may become a *coastal plain*. Such a plain will be covered with marine deposits, including silt, and is generally fertile. It may terminate against the former coast-line, perhaps in a line of cliffs. The best example of a coastal plain of this type is in the eastern United States, south of New York, where a plain between 50 and 200 kilometres wide has been formed. It

Sand dunes at Hengistbury Head (*F R Dobson*)

slopes gently to the sea and its western edge is the abrupt scarp known as the *Fall Line*, so called on account of the rapids and falls where the rivers cross it.

A subsidence of the land relative to sea level has a particularly marked effect on coastline in lowland areas. This can be seen in Suffolk and Essex, with their numerous shallow estuaries. Silting is taking place in the estuaries and sand spits have formed at the seaward end of many of them. In highland areas subsidence has produced rias or fiords (see Chapter 7). The silting process here is much slower since few large rivers flow into them. If there is a longitudinal coastline, with mountains running parallel to the coast, the tops of former ranges may become long, narrow islands, whilst former valleys become straits. This is often described as a Dalmatian type of coastline, from the Dalmatian coast of Yugoslavia.

Exercises

1 Which one of the following is a result of marine erosion? Spit, beach deposits, terrace, arch, moraine.

2 (*a*) Explain how and why a series of bays and headlands might develop along a stretch of coast.

(*b*) With the aid of labelled diagrams describe and explain the features formed as a headland is eroded.

3 Study the photograph on this page.

(*a*) What is the structure projecting into the sea?

(*b*) With the aid of a diagram or sketch plan describe and explain the purpose and effect of the structure.

(*c*) How have the sand dunes been formed?

(*d*) What suggests that some of the sand dunes are now fixed in place?

Figure 10.7 Coastal sand dunes

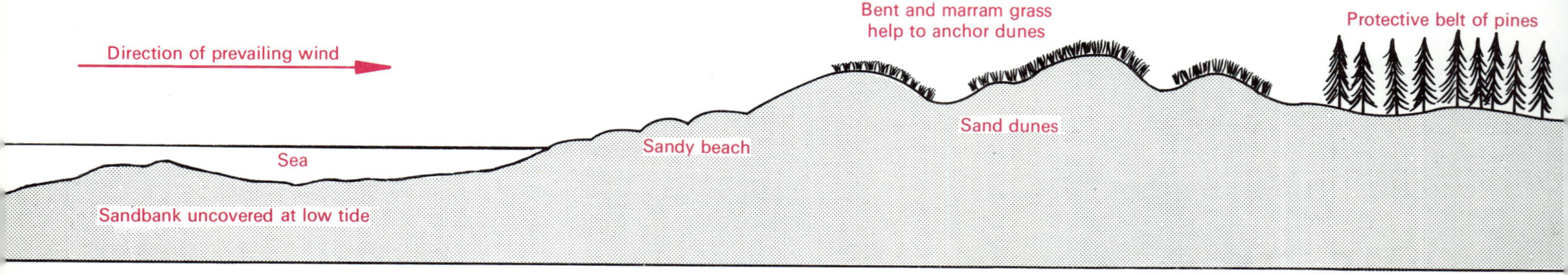

Chapter 11

Oceans and Seas

Although the oceans and seas cover about two-thirds of the Earth's surface the systematic study of the oceans was not begun until the years 1872 to 1876, during the round-the-world voyage of HMS *Challenger*, a ship specially equipped for oceanographic research. The discoveries made on this voyage concerning the depth, temperature and salinity of the oceans, the nature of the sea bed and the various forms of life in the sea laid the foundations of our knowledge of the physical geography of the oceans, often referred to as oceanography.

The Ocean Floor

Knowledge about the ocean floor has expanded enormously in recent years. The *ocean trenches* or *deeps*, situated mainly near coasts bordered by high mountains, indicate where an oceanic plate is diving beneath a continental plate. These ocean trenches are extremely deep, reaching over 11 000 metres below sea level in one case. They are thousands of kilometres long but very narrow and, as explained in Chapter 6, are linked with major zones of earthquake activity.

Much of the ocean floor consists of flat or gently sloping basins or *abyssal plains*, where sediment can accumulate in the form of very fine-grained ooze. *Oceanic rises* represent higher sections of the ocean floor. They occasionally break the surface, as with the Rockall rise in the North Atlantic.

Ocean ridges occur where new rock is being formed where ocean plates are moving apart. They are systems of mountain ranges rising up to 5000 metres above the ocean floor. The Mid-Atlantic Ridge is an example.

The Continental Shelf

A shallow shelf sloping gradually to the 200 metre line surrounds most land masses. This is the submerged part of the continents and is called the *continental shelf*. It is mainly covered by sand, mud and other deposits derived from the land and at its edge the sea bed falls steeply to the ocean depths. The shelf is very wide in some places, for example off the coast of north-west Europe, where the British Isles rise from it.

Temperature

Temperatures at the surface of the sea vary from about 27°C in tropical areas to freezing point in polar areas. The decrease does not take place at a uniform rate, and there are large variations even along the same line of latitude, the principal cause being the influence of ocean currents. Temperatures within the sea also decrease as one descends, and in deep waters are roughly the same for all latitudes. This constant value is about 4°C.

Salinity

On average there are 35 parts (by weight) of mineral matter in solution per 1000 parts of sea water. Rivers which carry dissolved minerals from the land must have contributed, since the land and sea were first formed, to this vast accumulation. By far the commonest mineral present in sea water is sodium chloride (common salt), in spite of the fact that there is very little of it in river water. On the other hand the large amount of calcium bicarbonate in river water contrasts strikingly with the low proportion in sea water. This is probably the result of its continual extraction by sea organisms in order to make their shells and skeletons. After the death of the organisms, their skeletons containing insoluble calcium carbonate are deposited on the sea bed.

There is a considerable variation in the salinity of the surface layers of the ocean in different parts of the world. In regions of low evaporation, heavy rainfall and where large rivers enter the sea the salinity is below average, whilst in regions of high evaporation, low rainfall and few rivers it is above average. These differences are accentuated in the case of partially enclosed seas where mixing with the ocean water does not readily take place. In the Baltic Sea, which is an example of the first type, the salinity is only about 8 per 1000, decreasing to 2 at the heads of the Gulfs of Bothnia and Finland. In the Mediterranean Sea, on the other hand, it is about 36.5 per 1000 in the west, increasing to 39 in the east. In the Red Sea it reaches 41 in the Gulf of Suez.

Figure 11.1 Section across the Atlantic Ocean from the West Indies to Denmark

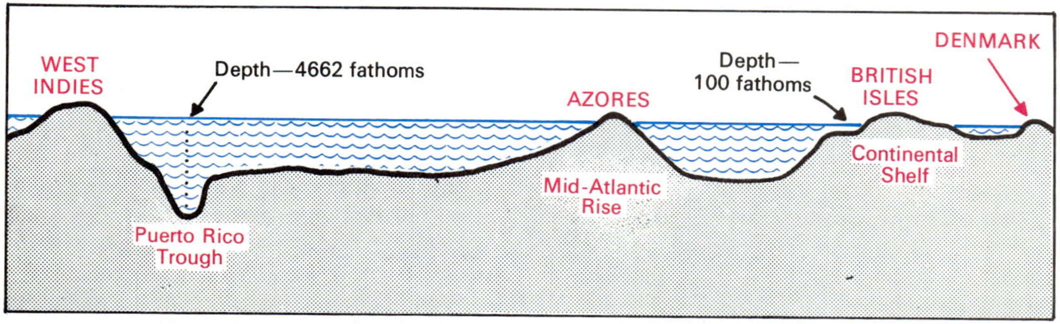

Ocean Currents

Surface movements of the ocean may take the form of *currents*, with speeds ranging from 2 to 10 kilometres per hour, or much broader, slower moving *drifts*. The principal causes of both currents and drifts are:

(1) variations in the density of the ocean water caused by differences of temperature and salinity (the differences of temperature give rise to a slow surface movement of warm water from the Equator to the Poles and a return movement of cold water beneath the surface),

(2) the effects of the prevailing winds, which blow the surface layers of water along with them, and

(3) the shape of the land masses, which alter the direction of many currents.

The general circulation resulting from these factors is well illustrated by the currents of the North Atlantic Ocean. The North and South Equatorial Currents flow from east to west, blown by the North-East and South-East Trades. Part of the South Equatorial Current is deflected northwards by the eastern corner of Brazil, so that it joins up with the North Equatorial Current. Some of this combined current enters the Caribbean Sea and Gulf of Mexico, the rest keeping to the east of the West Indies. The heaping up of warm water within the Gulf of Mexico gives rise to a powerful current which flows out of the Florida Strait and continues northwards, off the east coast of the United States, as the warm Gulf Stream. About Cape Hatteras the Stream passes into mid-Atlantic and spreads out to form the North Atlantic Drift, which is still warm for its latitude. On approaching the British Isles it divides, one part continuing to the coast of Norway and the other passing southwards as the relatively cool Canaries Current and eventually joining up with the North Equatorial Current. In the centre of the North Atlantic there is little surface movement and here is the Sargasso Sea, where the floating weed 'sargassum' is abundant, though there is no foundation for the belief that the weed can hinder the progress of ships.

In addition to this general clockwise circulation, a cold current called the Labrador Current flows southwards from the Arctic. It hugs the American coastline as far south as Cape Cod, where it gradually sinks beneath the warm waters of the Gulf Stream. The southerly drift of icebergs to Newfoundland and beyond is one effect of this current.

The general circulation in the South Atlantic resembles that of the North Atlantic except that it is anti-clockwise, and at its southern end, where the South Atlantic opens out into the Southern Ocean, it is affected by the cold West Wind Drift. The

Figure 11.2 Main ocean currents ▶

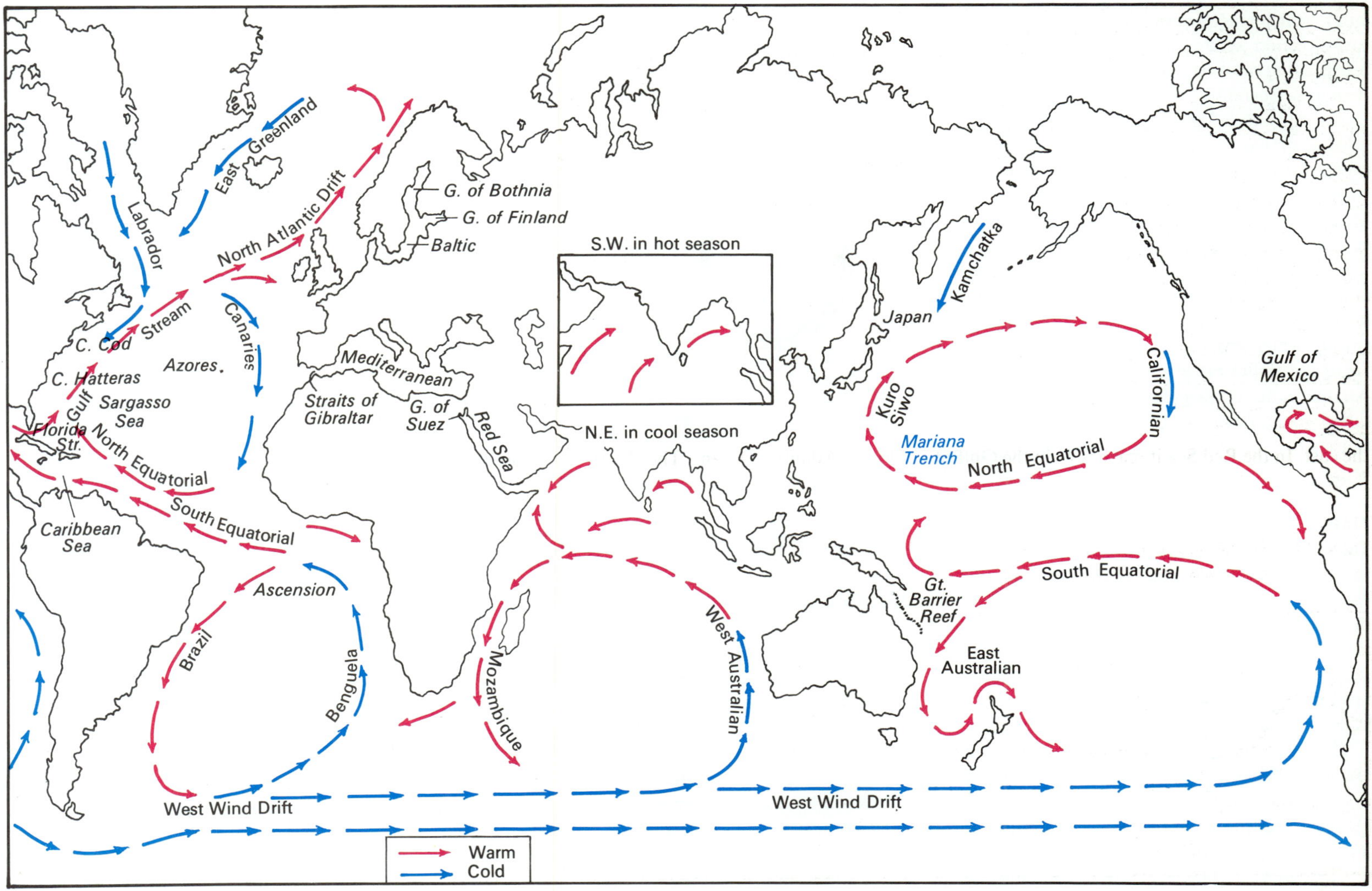

East Greenland
Labrador
North Atlantic Drift
G. of Bothnia
G. of Finland
Baltic
S.W. in hot season
Kamchatka
Japan
C. Cod
Stream
Canaries
Azores.
C. Hatteras
Gulf
Sargasso Sea
Mediterranean
Straits of Gibraltar
G. of Suez
Red Sea
N.E. in cool season
Kuro Siwo
Californian
Gulf of Mexico
Mariana Trench
North Equatorial
Florida Str.
North Equatorial
South Equatorial
Caribbean Sea
Ascension
Gt. Barrier Reef
South Equatorial
Brazil
Benguela
Mozambique
West Australian
East Australian
West Wind Drift
West Wind Drift
Warm
Cold

latter, which travels round the world in a west to east direction, sends a tongue of cold water northwards along the west coast of Africa, known as the Benguela Current.

The currents of the Pacific and Indian Oceans have a similar pattern to those of the Atlantic except for the northern part of the Indian Ocean, where the seasonal changes known as the Monsoons produce a change in the direction of the currents, as shown in figure 11.2.

In addition to these ocean currents there are numerous local currents. Some of these are caused by differences of salinity. An example is the surface current which flows from the Atlantic through the Straits of Gibraltar into the Mediterranean. The Atlantic water is less saline and is replaced by more saline water from the Mediterranean which, because of its greater density, moves along the sea bed.

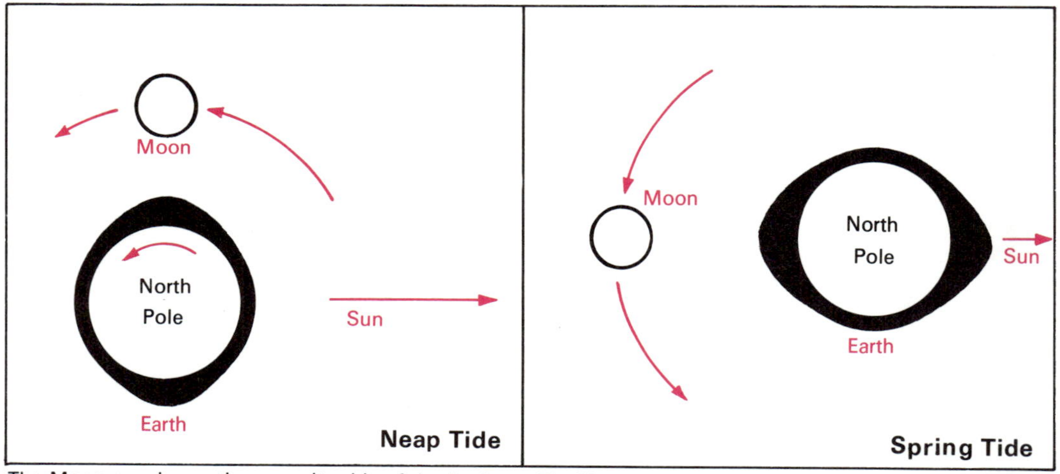

The Moon may be on the opposite side of the Earth

The Moon may be on the same side of the Earth as the Sun

Figure 11.3 Tides

Tides

Tides are caused by the gravitational pull of the Moon and to a lesser extent of the Sun, resulting in two great bulges in the oceans on opposite sides of the Earth and in line with the Moon. These bulges move round the Earth as it rotates, giving two tides each day at any one place. Since the Moon moves in orbit in the same direction as the Earth rotates, the high tide is about 52 minutes later each day.

The difference between the levels of the high and low tides, which is known as the *tidal range*, does not exceed 1 metre in mid-ocean, but when the tidal bulge reaches the coast the water piles up, producing tidal ranges normally between 5 and 7 metres, but exceeding 12 metres in certain funnel-shaped

7 metres, but exceeding 12 metres in certain fun-
estuaries. Some partially enclosed seas, including the Mediterranean and the Baltic, have a very small tidal range.

When the Sun, Moon and Earth are in a straight line the gravitational pull of the Sun and Moon work together, causing especially high tides, known as *spring tides*, but when the Sun, Earth and Moon form a right angle low tides, called *neap tides*, are experienced. A study of figure 3.2 (the 'Phases of the Moon') will show that both the spring and neap tides occur twice a month, the spring tides with the new Moon and full Moon, and the neap tides with the first and last quarters.

Life in the Oceans

Probably the most important type of life in the oceans and seas is known as *plankton*. It consists of microscopic forms of plant and animal life with only limited powers of movement and largely at the mercy of tides and currents. The plant plankton is fed on by the animal plankton which in turn form the food of fish and crustacea.

Cold seas are richer in nutritive elements, including nitrates and phosphates, than warmer seas. This leads to a greater abundance there of plankton, fish and larger forms of life such as whales, seals and penguins. The world's richest fishing grounds, however, are off the coasts of north-west

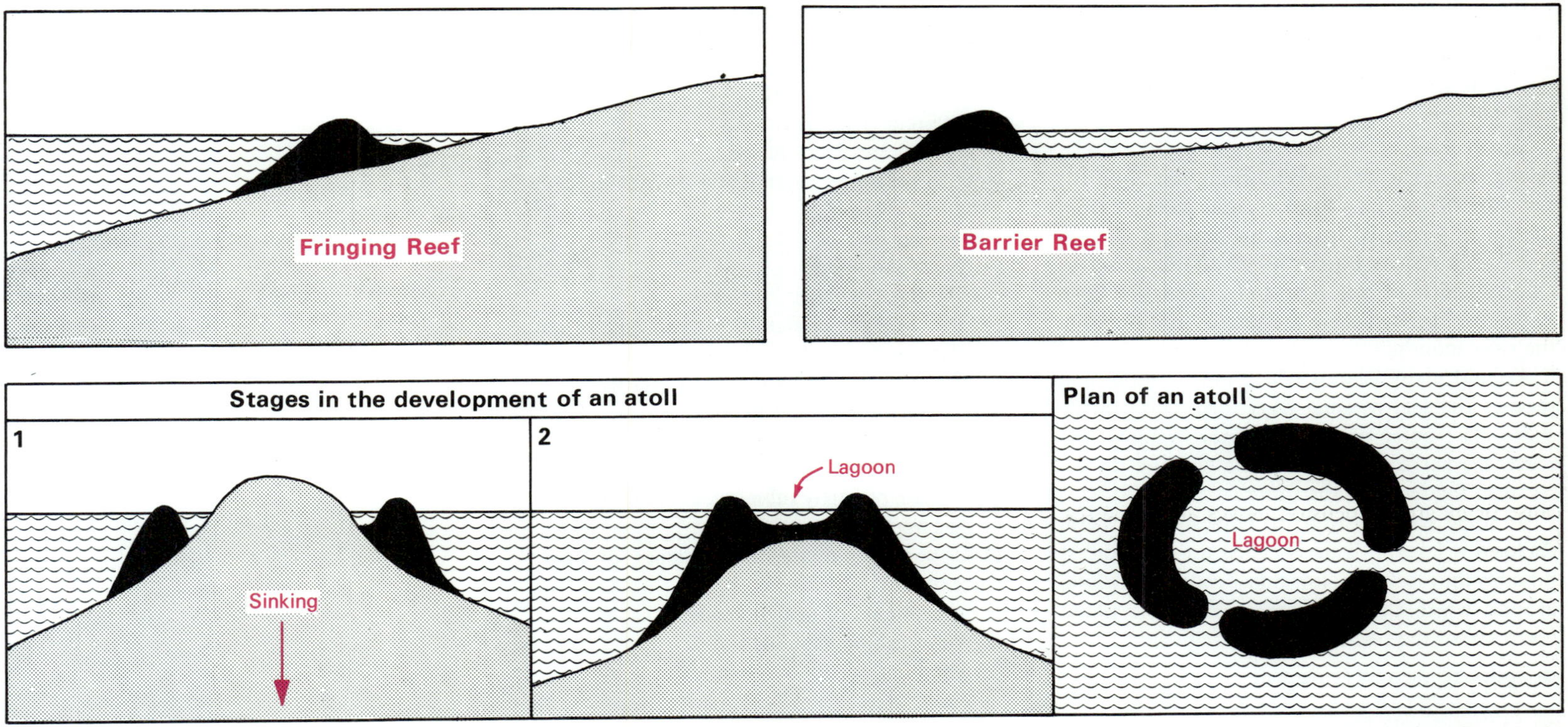

Figure 11.4 Coral reefs

Europe, eastern Canada and Japan. These areas lie (*a*) where there is a broad continental shelf, since many fish live, feed and lay their eggs on the sea bed, (*b*) at the meeting and mixing place of currents, which bring nutritive elements and plankton, and (*c*) where there is considerable wave action and tidal range to aerate the water.

Coral Reefs

A *coral polyp* is a small creature with a skeleton and base of calcium carbonate, the base being attached to the sea bed. When it dies the soft part of its body quickly decays or is washed away, leaving the limey base and skeleton behind. Where the corals live in colonies their remains fuse together to form hard rock. New corals grow on top of this rock, on which in turn they leave their skeletons, so that the coral is built up to form a reef.

Corals live only in warm waters and cannot stand temperatures below 20°C, so that they are mainly restricted to the belt between latitudes 30° north and south, though they are found around Bermuda (32° north) because of the warming effect of the Gulf Stream. Coral will die when it is exposed to the air for long periods so that the reef ceases to grow upwards when it reaches sea level. It is found at depths of less than 60 metres and the sea must be free from sediment.

Types of Reef

Along coastlines where rivers bring mud into the sea, *barrier reefs* may form at some distance from the coast, separated from it by a stretch of calm water. The Great Barrier Reef off the coast of Queensland is of this type and extends for 2000 kilometres, though there are breaks in it opposite the river mouths.

Around many islands where only a little mud reaches the sea, *fringing reefs* form close to the shore. The outer corals grow more rapidly and reach the surface first, leaving a shallow lagoon between the reef and the shore.

In some cases a ring or horseshoe of coral encloses a lagoon. This is called an *atoll* and is possibly the result of a slow subsidence of the island around which the coral grew. When the island disappeared the reef remained.

disappeared the reef remained.

Exercises

1 Trace an outline map of the Atlantic Ocean. Mark on it and name
   (*a*) the continental shelf,
   the mid-Atlantic ridge,
   the Puerto Rico trench,
   (*b*) the following islands on the continental shelf:
   British Isles,
   Newfoundland,
   Falkland Islands,
   (*c*) the following islands of volcanic origin: Azores, Cape Verde Islands, Ascension, St Helena, Tristan da Cunha, Canaries,
   (*d*) the following coral islands:
   the Bahamas,
   Bermuda,
   (*e*) a chain of partly submerged fold mountains: the West Indies.

2 Describe and explain the influence of ocean currents on
   (*a*) the climate of nearby lands,
   (*b*) large fishing grounds,
   (*c*) shipping.

Savanna in Kenya (*J Allan Cash*) ▶

PART FOUR

The Natural Regions of the World

Chapter 12

Equatorial Regions

Climate

In lowland areas between 8° north and 8° south of the Equator it is hot and humid all the year round. Singapore may be taken as typical of these regions. The temperature graph in figure 12.1 shows it to have an average of about 26°C, with very little variation from month to month. Rain falls throughout the year and amounts to 2410 millimetres. It is mainly of a convectional type, resulting from the ascent of hot air within the equatorial low-pressure belt (the Doldrums).

Weather conditions vary little from day to day. At about 6 am the Sun rises and soon disperses the early morning mists. As it climbs higher in the sky temperatures rise rapidly, exceeding 32°C by noon. In the afternoon clouds begin to form, developing into thick masses of cumulonimbus and culminating in a heavy downpour of rain, often accompanied by thunder. Afterwards the clouds break up and the Sun sets at about 6 pm, followed by a mainly clear night.

Natural Vegetation

On these lowlands grows a dense, luxuriant forest of broad-leaved evergreen trees, some of which attain heights of up to 60 metres. As a result of the uniform climate there is no particular season for flowering and fruiting. The taller trees struggle upwards towards the light and their upper branches spread out to form a canopy, whilst below, in the damp, oppressive twilight, is a tangled undergrowth of palms, bushes and creepers. There is a great variety of trees, most of which are *hardwoods*. In shallow coastal waters, particularly around deltas and estuaries, and along the rivers, a close mass of small trees grows from the mud, forming what are called *mangrove swamps*.
swamps.

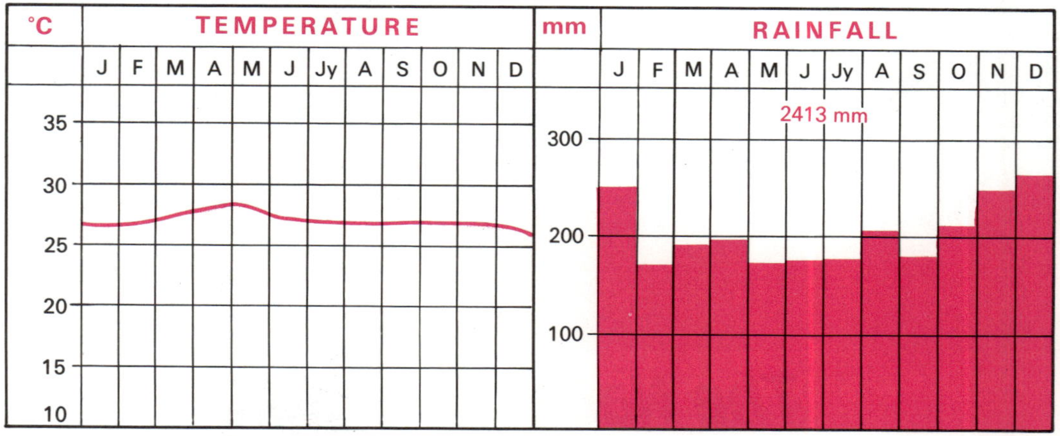

Figure 12.1 Climate graphs for Singapore

Mangrove swamps in the equatorial region of Nigeria What problems can this vegetation form provide in the human development of this area? (*Photo by Jack Barker, courtesy of the United Africa Co Ltd*)

THE AMAZON BASIN

The River Amazon and its tributaries drain a vast lowland area which stretches over 3000 kilometres inland and is in places 1500 kilometres across. A thick deposit of alluvium covers the lowland and provides fertile soils which are enriched by decaying vegetation. An equatorial forest of extraordinary density, known as the *selvas*, covers most of this plain.

The extensive swamps, the hot sticky conditions and the mosquitoes, some of which carry malaria and yellow fever, have all proved a barrier to economic development. Consequently the Amazon Indians remained largely unaffected by outside influences until relatively recently, particularly those living in western regions and areas far away from the main waterways. They live in small clearings in the forest, which have been made either by axe or by burning. The clearings are at some distance from the rivers, partly to be above the flood level and partly to avoid insects which are most numerous near the river banks. The huts are constructed of local materials and consist of a timber framework lashed with creepers, folded palm leaves covering the roof. Sometimes the tribe joins in constructing a single large communal hut.

Figure 12.2 Tropical regions

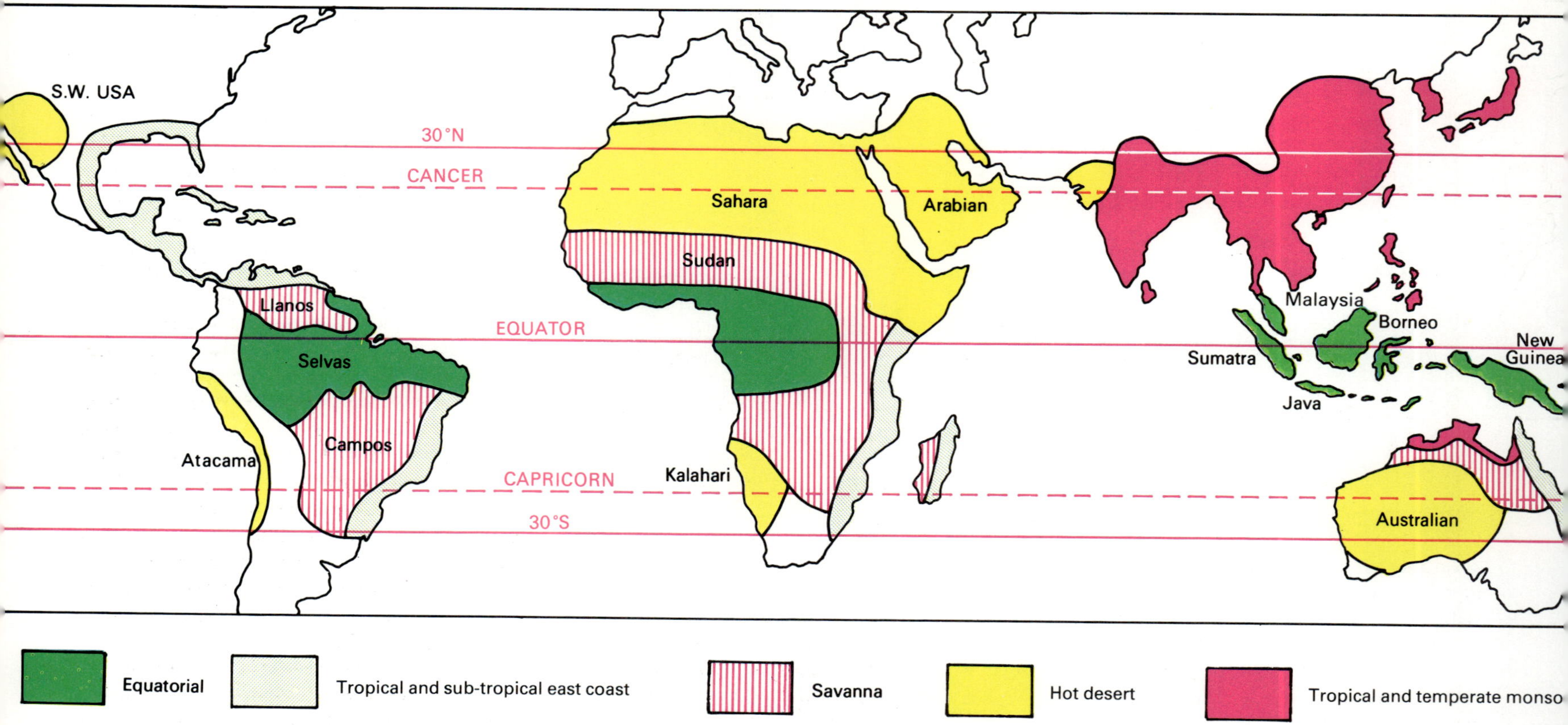

Figure 12.3 The Amazon basin

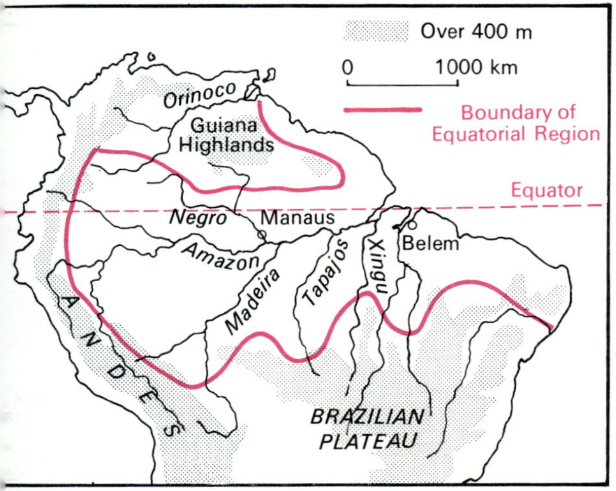

Around the huts the land is divided into garden plots. *Manioc*, which is the most important crop, is a shrub with large tuberous roots which look and taste rather like parsnips. The root is peeled and then the inside is ground into flour and baked into a leathery type of bread. Yams and sweet potatoes are also grown.

After three or four years the clearing is abandoned and the whole village moves off to make a new clearing, with new huts and gardens. This method of *shifting tillage* is necessary because the plant nutrients released by burning cleared vegetation are soon used up or washed out of the soil by heavy rains.

Besides being cultivators the Amazon Indians are hunters and fishermen. Their prey includes tapir, rodents, small monkeys, birds and reptiles as well as a large variety of fish.

The crafts of these people reflect their forest environment. Clothing is frequently made from bark which has been pounded to make it pliable, and creepers are used to make baskets.

Communications have always been a problem, but the rivers are much used for transport, the principal ports being Manaus and Belem. The Brazilian government has decided to open up the Amazon forest region to settlement and is constructing a network of roads. The most important of these is the Trans-Amazonica Highway. New farming areas have been established along the roads, and the idea was that this area would provide a home for people from the north-east of Brazil which is densely populated and subject to frequent droughts. Large companies have also taken over huge areas of forest to clear for grazing land for beef cattle. However, within ten years some of the areas converted to grazing land have had to be abandoned. Plantations have also been developed in some areas, but, again, not all have been successful.

The effect on the Amazon Indians has been catastrophic. Partly as a result of diseases introduced from outside, but largely due to the destruction of forest, the original tribal Indian population of 6 to 9 million has been reduced to less than a quarter of a million.

The forest is a rich source of products useful as food, raw materials and medicines, but exploration is still so limited that very many plants have not yet been identified. The forests have already provided many species of plants now widely cultivated, such as rubber and coffee, and it is thought that there are very many more still undiscovered.

EQUATORIAL AFRICA

The equatorial lands of Africa cover the Zaire basin and most of the coastal lowlands of West Africa. Natural conditions are very similar to those of the Amazon, but many of the inhabitants are more advanced than the Amazon tribes and there has been more influence by European colonial powers. In the more accessible areas the cultivation of commercial crops has been developed, and of these cocoa and palm oil are outstanding.

Cocoa is a typical crop of hot, wet forest areas and requires constant heat and moisture as well as shelter from strong winds. The cacao tree from which cocoa is obtained grows to about 6 or 9 metres in height, and the pods form on the main stem and larger branches. The cocoa farms in West Africa are normally family concerns growing on the one hand a money crop, cocoa, and on the other hand food crops such as yams, cassava (the African variety of manioc), plantains and vegetables. The farm is also likely to have livestock including chickens, goats and pigs.

Cacao trees do not bear fruit until four or five years old. There are two harvests in the year, the main one between October and December and a smaller one between March and May. The pods are cut from the trees and split; the beans are then collected in baskets and piled on plantain leaves with more leaves put over as a cover. They are then left for a week to ferment, which improves the flavour and keeping quality. Afterwards they are dried in the hot sun for five or six days, put into sacks and taken by the farmer to the local marketing centre from where they are taken to a port and shipped abroad, probably to Europe or the United States. The chief exporting country is the Ivory Coast.

The final preparation of cocoa takes place in the importing country. Here the beans are first graded and cleaned, then roasted and ground. To make cocoa, some of the cocoa butter is pressed out and the residue crushed to powder. To make chocolate, the cocoa butter is retained and sugar added before the moulding process.

The *oil palm* grows wild in the forests. It is also cultivated on plantations owned by large companies, and on small farms. In some countries, such as Zaire, plantation production is more important, but in others, such as Nigeria, small-farm production is more important.

The pulp which surrounds the kernel provides an oil which can be used for soap, candles and grease. Large quantities of the kernel are exported, especially from Nigeria, to European and North American factories where it is crushed to provide a valuable edible oil used in making margarine.

The rain forests of the equatorial region of Africa have been developed for their reserves of hardwood timber to a far greater extent than those of the Amazon basin. Since 1950 Africa as a whole has lost 52 % of its forests, and most of this loss has been in rain forest regions. In individual countries the rate of loss has been exceptionally high. In the Ivory Coast, 12 million hectares of rain forest have been reduced to less than 4 million within 25 years. Cutting is continuing at a rate of 500 000 hectares each year.

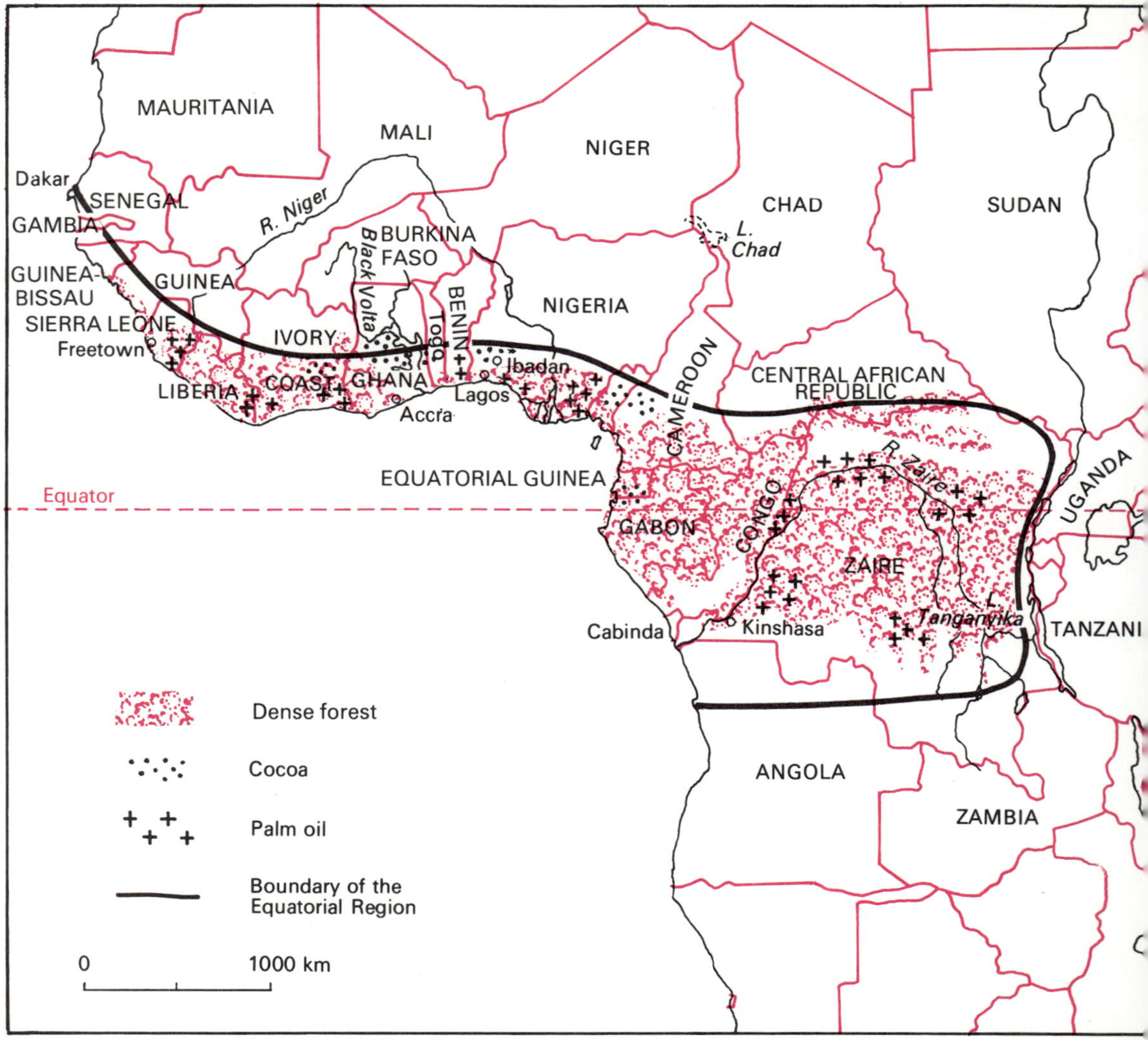

The third equatorial region comprises Malaysia and the East Indies. The high temperatures here are partly mitigated by sea breezes and by the mountainous nature of much of the land. There is great diversity among the islands, in their peoples' culture, religion, degree of industrialisation and so on. Access is easy by sea and air. Good communication links have helped the region to develop. In the nineteenth century it was colonised by Europeans for its resources. Later, many Chinese and Indians also settled in these lands. The legacy of the colonial powers, especially the British in Malaysia and the Dutch in Java and Sumatra, is visible in many areas of economic life. For a long time the region was noted for its spices. Then came the establishment of rubber plantations in Malaysia and Indonesia in the late 1800s. Now there is a varied agriculture including the growing of rice, sugar cane, tea, coffee and cinchona.

Rubber is a native plant of the Amazon basin, and seeds were taken from there to the hothouses at Kew Gardens, London. From here the seedlings were transferred to plantations in Malaysia and Sri Lanka. It was not until 1910 that Malaysian rubber entered world markets in considerable quantities.

The rubber tree normally grows to a height of about 20 metres and will probably yield about two kilograms of rubber a year. If it remains healthy it will continue to do so for over thirty years. Tapping, which begins when the tree is about five years old, is a skilled process with the object of severing the latex tubes in the inner bark. The latex is collected in cups and taken to the plantation

Figure 12.5 The manufacture of rubber

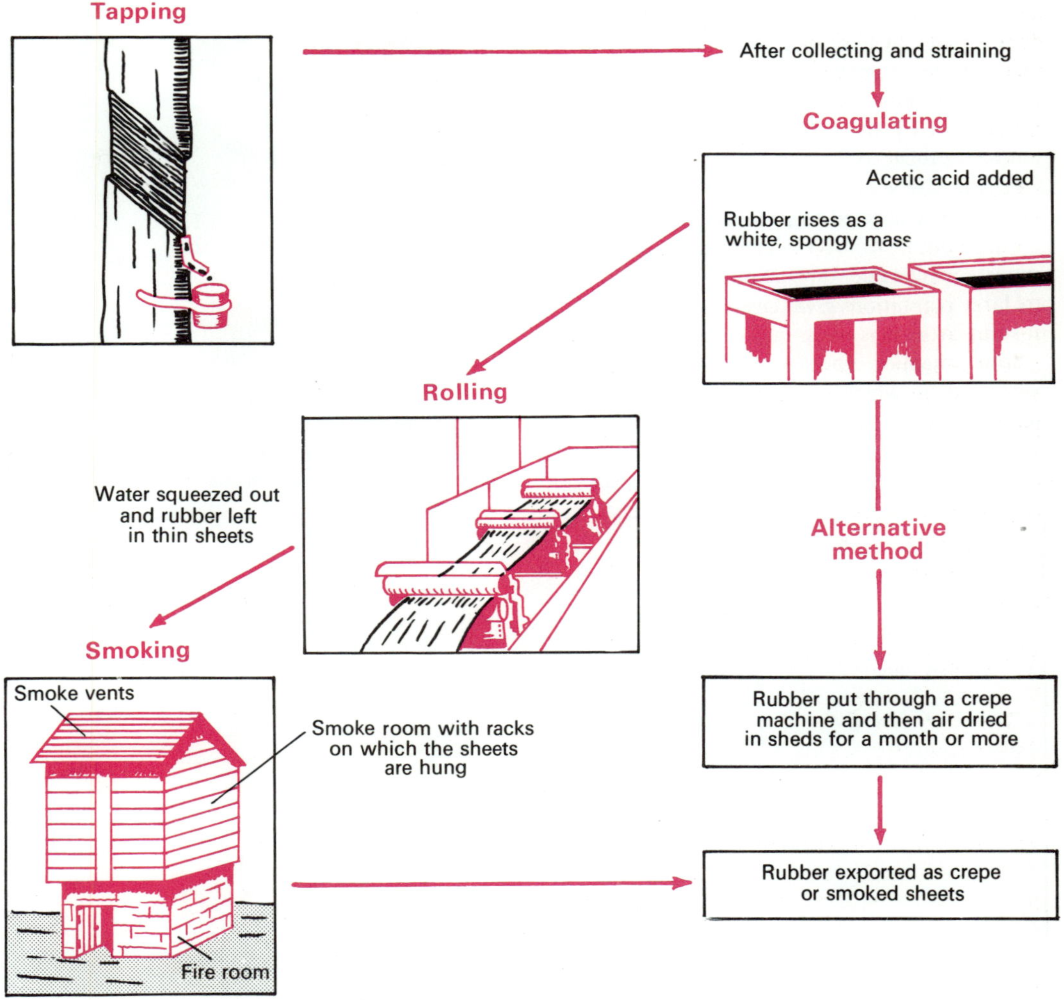

factory where it is sieved and diluted. Acid is then added to coagulate the latex which is pressed and dried, either in a smoke room to form smoked sheets or in the open to form crepe. It is normally exported in this condition, and other processes are carried out by the importers. The main uses of smoked rubber are in motor car tyres and in footwear. Crepe rubber is also used in footwear, as well as being aerated to produce foam rubber.

Palm oil is a crop which has increased in importance considerably in South-East Asia. There are large plantations given over to the oil palm and in some areas it is more important than rubber.

Some mention should also be made of the *coconut palm* which grows in coastal areas. The nuts, which are surrounded by a fibre called *coir* (used for making matting and ropes), are split open and the fleshy part removed to be dried in the sun. The product is called *copra* and yields valuable coconut oil when crushed.

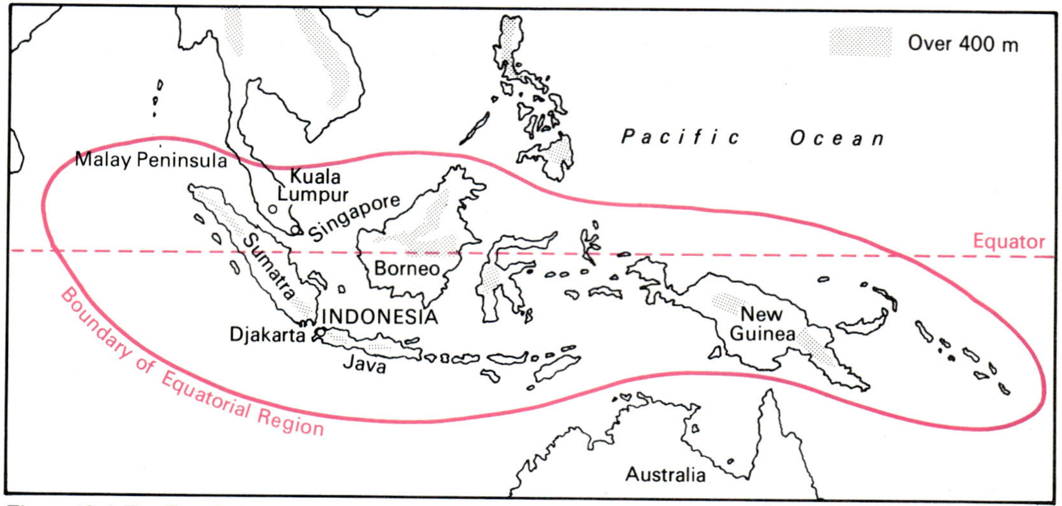

Figure 12.6 The East Indies and Malaysia

Harvesting cocoa beans (*Cadbury Schweppes Ltd*)
Describe the scene and then explain the various processes which take place before the beans are sent to market.

Describe the conditions of growth and the uses of the palm oil illustrated here (*Photo Unilever Ltd*)

Loading logs in Ghana
What are the problems facing the lumbering industry in equatorial West Africa? (*Photo United Africa Co Ltd*)

◀ **Rubber tapping** (*Dunlop Ltd*)
Describe the activities taking place.

Exercises

1 Answer the following questions, referring back to Chapter 4 where necessary:
  (*a*) Why are there uniformly high temperatures in the equatorial region?
  (*b*) Why does most of the rain fall in the afternoon?
  (*c*) On what dates is the Sun vertically overhead on the Equator, and what connection is there between this and the absence of seasons?
  (*d*) Why are the trees in the equatorial region broad leaved and evergreen?

2 Explain why a country like Ghana, with 66% of its exports made up of one primary product, cocoa, can be greatly affected by changes in world trade.

Chapter 13

Savanna Regions

Climate

The seasonal movement of the Doldrums, following the overhead sun, has already been referred to in Chapter 3. This brings rain during the hot season to the savanna regions, which lie between latitudes 8° and 18° north and south. The cool season is practically rainless as these areas are dominated by the Trade winds which, having travelled over land, are dry.

The climatic statistics for Kano, in a savanna area in north Nigeria, show a hot season extending from March to October with mean temperatures over 27°C, but also a slight lowering of temperatures during the period of heaviest rain. The cool season has mean temperatures of between 20°C and 25°C. The rainfall of about 875 millimetres, occurring mostly between June and September, is normal for the savanna, although towards the equatorial areas there may be up to 1500 millimetres, whilst towards desert regions it may be below 500 millimetres.

Natural Vegetation

The drought prevents the growth of luxuriant forest; instead tall grasses with scattered, stunted trees and thorn scrub are found. The trees are adapted to overcome the drought by having few and small leaves and thick bark so that the loss of moisture by transpiration is restricted. The acacia and the weird baobab tree with its vast spongy trunk capable of storing water are typical sights in many parts of the African savanna. Most trees are deciduous, shedding their leaves during the cool dry season. The grass withers and dies down during the dry season and is then often burnt to improve the next season's growth. When the rains break the grass sprouts afresh and grows rapidly. Some varieties including the elephant grass grow to over 3 metres in height.

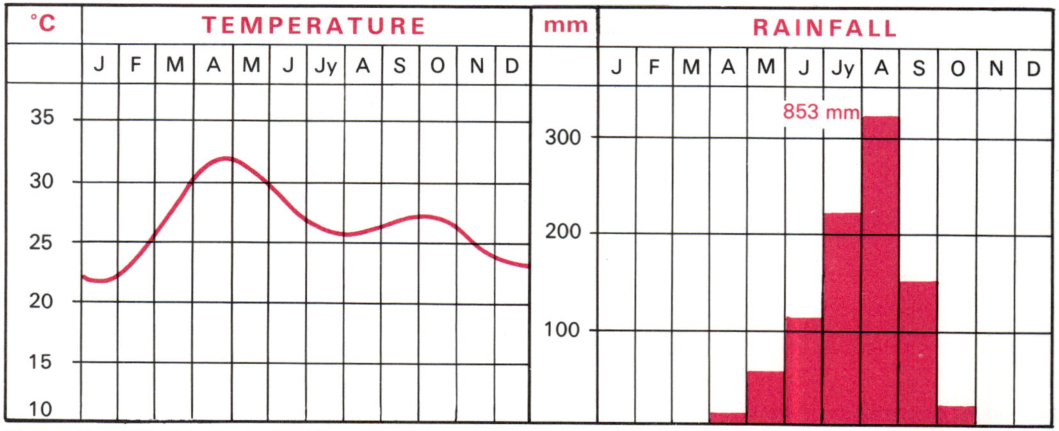

Figure 13.1 Climate graphs for Kano

Kano River; dry season – wet season (*S Arnold*)

THE SAVANNA REGIONS OF AFRICA

On a journey from the Guinea coast of West Africa northwards the change from a region of equatorial rain-forest to one of tropical grassland is very gradual. First the forest begins to thin and natural clearings appear, which become more extensive as one travels northwards until eventually open country with tall grass and scattered trees is reached. Farther north, about 800 kilometres from the coast, the grass becomes coarser with the decrease in rainfall, and as one approaches the Sahara Desert it grows in tufts, together with many thorn bushes. This grassland stretches from Cape Verde on the Atlantic to the Ethiopian Highlands and is called the Sudan. Except for coastal areas, similar conditions exist to the east of the Zaire rain-forests in Kenya, Uganda and Tanzania, and also in the south in Zambia, Zimbabwe and Angola.

These regions have always been noted for their great variety of wild animal life. The two main groups are the swift-footed, vegetable-eating animals such as the antelope, zebra and giraffe, and the flesh-eating species including the lion, leopard and cheetah. In recent years, however, their numbers have declined rapidly in spite of game reserves in various parts of the region, and some species are in danger of extinction as people take more and more land for farming.

Cattle rearing is carried on extensively on the grasslands, although nearly everywhere this activity is combined with some form of soil cultivation. The cattle are generally of the humped *zebu* type and their quality has for long been poor. The drought of the cool season which brings shortages of water and pasture makes the animals weak and susceptible to disease. The *tsetse fly* brings disease to livestock and has discouraged cattle rearing over large areas of the wetter savanna. Although in most areas only the hides are of value for export purposes, there is a small but growing export of canned beef, particularly from East Africa.

Besides his cattle the typical African farmer has his own field which he cultivates with the help of his family. At the end of the dry season the weeds are burned and this helps to fertilize the land. The field is broken up, usually with a short-handled hoe, and the crops grown include maize, millet, guinea corn, cassava, yams, beans, gourds and bananas. *Maize* and *millet* are the principal food crops. Maize, which thrives on the more fertile and better-watered land, was introduced by the Portuguese from America. Millet is grown more in the drier parts.

Together with these the farmer frequently grows a cash crop. This is a crop which is grown for sale, normally for export. An example is *groundnuts* which provides a valuable oil used in margarine and soap manufacture and also foliage for use locally for fodder.

Particularly in parts of East Africa the cultivation of commercial crops on large plantations is widespread. Sugar cane, cotton, tobacco and sisal (a fibre used for making rope) are all grown on plantations as well as on smallholdings.

The drier parts of the savanna regions are the areas most susceptible to desertification, which has been described in Chapter 8. In these areas rainfall is erratic and severe droughts have occurred throughout the 1970s and 1980s in the zone to the south of the Sahara desert known as the Sahel.

Figure 13.2 Savanna areas of Africa and South America

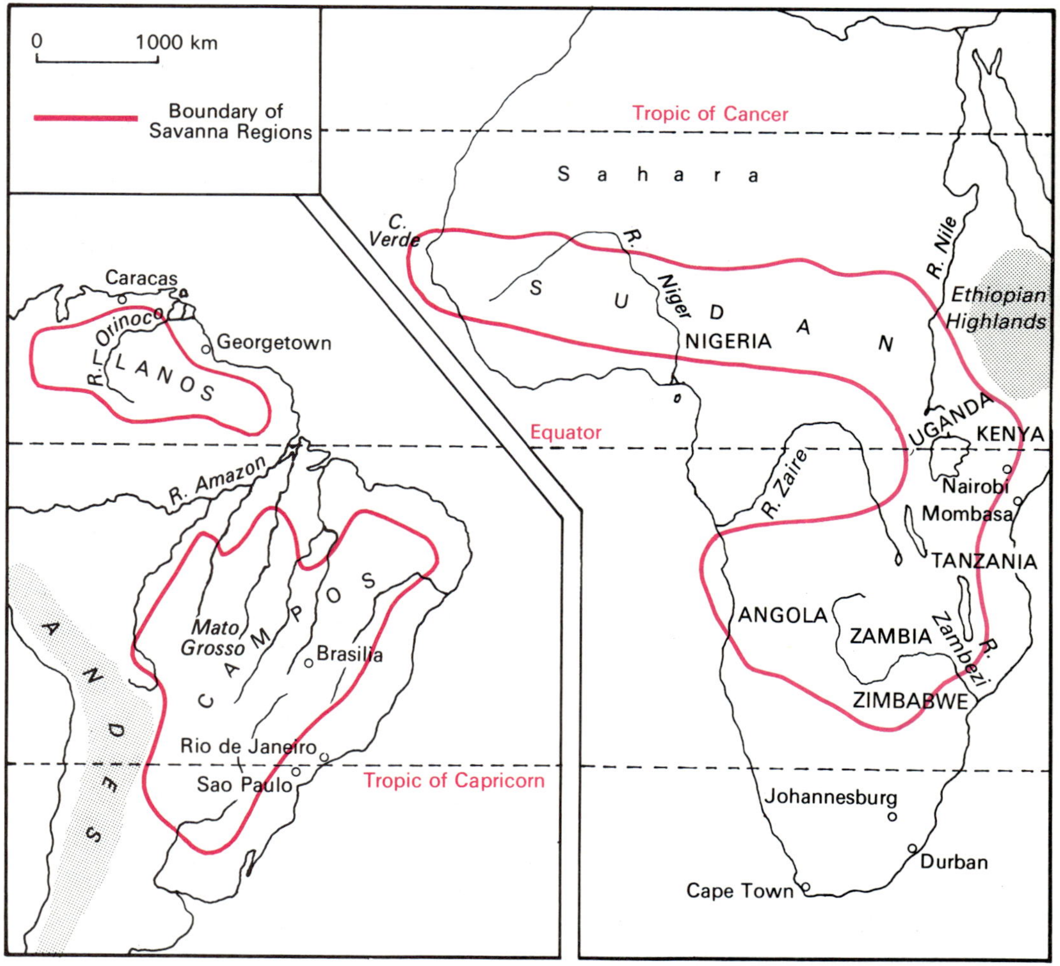

Baobab (*S Arnold*)
How is the baobab tree adapted to a savanna climate?

top right **Harvesting groundnuts** (*Unilever Ltd*)
What are groundnuts used for, and under what conditions will they grow? Explain why groundnuts are harvested in this way.

bottom right **African Market, Blantyre** (*C Seath*)
Where is Blantyre? What European and what African influences can you see in this scene?

THE SAVANNA REGIONS OF SOUTH AMERICA

The Llanos

This area lies in Venezuela and is the drainage basin of the River Orinoco. It has been little developed apart from cattle rearing. The Orinoco is an unsatisfactory means of communication since there is a great seasonal variation in flow. After the rains the low-lying grasslands are frequently inundated, making stock rearing near rivers particularly hazardous, whilst away from the rivers much of the pasture is poor. Consequently the quality of the cattle is low and their chief value at present is in their hides.

The Campos

This is the vast savanna area of Brazil. It lies on a plateau which is deeply dissected by river valleys. Cattle rearing is developing on the more accessible parts of the plateau, but vast stretches of the interior, including the Matto Grosso, remain largely undeveloped. The importance which the Brazilian Government attaches to the campos, partly on account of its great mineral wealth, is reflected in their building of Brasilia, the country's capital, in this region.

THE SAVANNA AREAS OF AUSTRALIA

The population of Australia's tropical grasslands is very sparse. A single cattle station is likely to cover hundreds of square kilometres. The lack of good communications and periodic droughts are handicaps to development.

Water supply in the dry season often depends on artesian bore holes. Marketing the beef presents great problems and many cattle are still driven along stock routes to the coast or the nearest railhead. Road trains, drawn by huge lorries, are now more common.

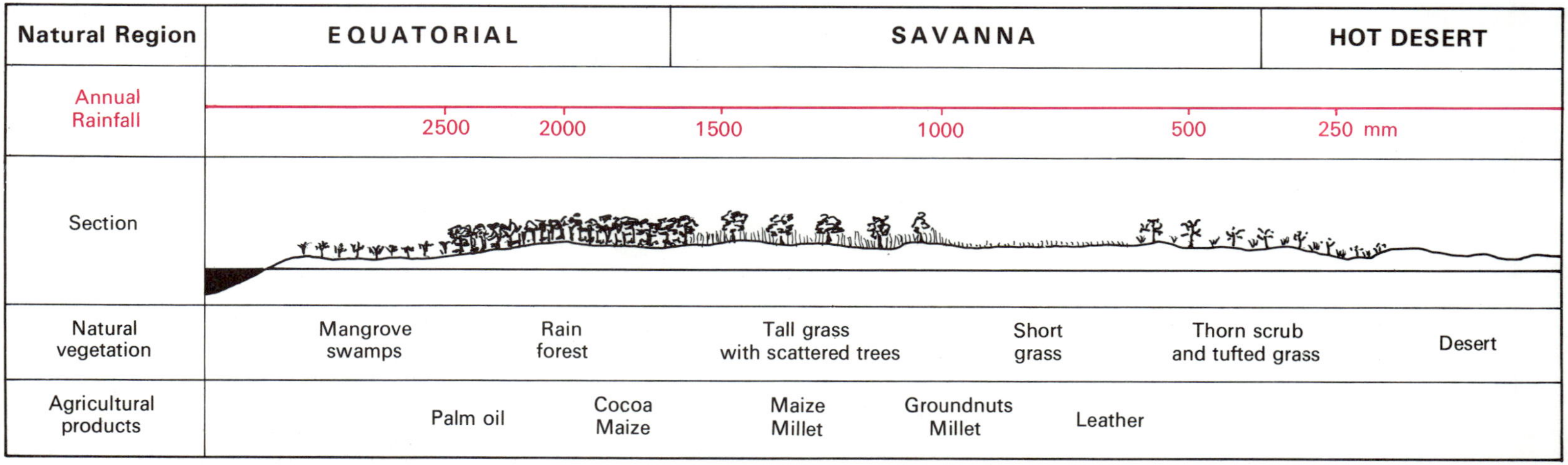

Figure 13.3 A section from the Guinea Coast to the Sahara

Write an account of a journey from Accra to Timbuktu. Describe the changes in scenery you would expect to find and say why these changes occur.

THE TROPICAL AND SUBTROPICAL EAST COASTS

These regions are very similar to the savanna areas, the chief difference being one of rainfall distribution and amount. The Trade winds here blow from the sea to the land and consequently produce heavier and more reliable rainfall, fairly well distributed throughout the year. Because of this the natural vegetation is more luxuriant and a greater variety of crops can be grown. Two crops, coffee and sugar cane, are particularly important.

Coffee is grown mainly in the state of São Paulo in south-east Brazil, on large estates called fazendas. The coffee plant is kept pruned to the size of a small shrub about $2\frac{1}{2}$ metres high. It has small shiny leaves and produces a berry which is purple in colour when ripe. The berries are picked, soaked and then mashed so that the pulp can be washed away to leave the beans. After drying, rollers are used to break off the husks; then the beans are graded and bagged for export. The overproduction of coffee between 1900 and 1930 has led to the introduction of other commercial crops which now include oranges, cotton, sugar cane, tobacco and bananas. Apart from Brazil, coffee growing has spread to many other countries, especially Kenya and Uganda in East Africa.

Sugar cane flourishes on the rich, deep soils of alluvial coastlands. The West Indies, particularly Cuba, and Central America are the leading producing areas in this region. The canes are grown from cuttings and take from 8 to 15 months to reach a height of 6 metres, when they are ready for cutting. Sugar cane requires a heavy rainfall and constant high temperatures, but a dry period just before harvesting is necessary to ensure that the juice is not too watery.

Considerable labour is required for cutting the cane and it is very hard work. The canes are then taken to a mill where they are crushed so that a brown liquid is extracted; this is filtered and then heated to leave a residue of unrefined sugar crystals. Large quantities of unrefined sugar are exported, and the refining takes place in the importing country. This involves further dissolving and filtering.

Harvesting sugar cane
(*Tate and Lyle Ltd*)

One final crop that is particularly associated with tropical east coasts is the *banana*. It is best suited to an environment similar to that required by sugar cane. The keynote of production is the rapid and careful marketing of the fruit so that it arrives at its destination not over-ripe.

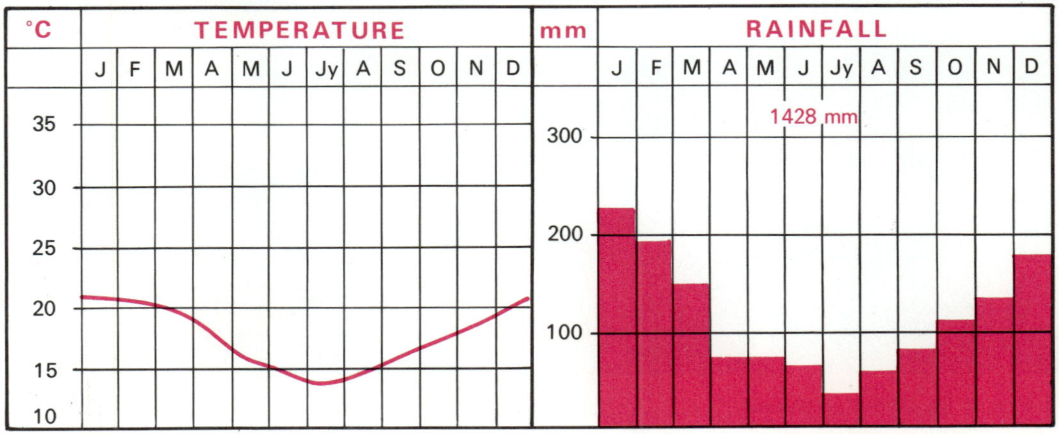

Figure 13.4 Climate graphs for São Paulo

Village in the Celebes (*Aerofilms*)

Bida, central Nigeria (*S Arnold*)

Exercises

1 Answer the following questions. Refer to earlier chapters where necessary.
  (*a*) Why does the rainfall in the savanna come almost entirely in the hot season?
  (*b*) Explain how the vegetation of the savanna is adapted to withstand a long drought.
  (*c*) How does the natural vegetation differ *within* the savanna region?

2 Draw a carefully labelled sketch map based on the following description of a coffee fazenda:

'The planter's house lies on the valley floor about 700 metres above sea level. Adjacent to it are stables for the horses and garages for the lorries. Near by also are extensive concrete yards for drying the beans in the hot sun, and the factory where the beans are hulled and sorted. A short distance away are rows of small bungalows for the workers, surrounded by fields on which they grow much of their food, including maize, potatoes and beans. All around on the valley sides at heights of between 800 metres and 1600 metres are the coffee bushes, thousands of hectares of them. There are special nursery beds for the coffee seedlings, and roadways linking up the various parts of the fazenda. A narrow-gauge railway runs from the factory to the local town.'

Old part of Kano, northern Nigeria (*S Arnold*)
Compare the homesteads illustrated. Explain how each reflects its natural environment.

Chapter 14

Hot Deserts

Climate

The hot deserts lie on the western sides of land masses between latitudes 15° and 30° north and south. The North-East and South-East Trades which are found in these latitudes have passed some way over the continents and are dry by the time they reach the deserts. For this reason the total annual rainfall is below 250 millimetres and over great stretches it is below 25 millimetres. The rain occurs in sporadic showers and is soon soaked up or evaporated.

Temperatures are affected by the almost complete absence of cloud and vegetation so that much of the Sun's fierce heat is absorbed by the rock, stones and sand which form the desert surface. During the hot season air temperatures frequently climb by day to beyond 45°C; on the other hand by night there is a great loss of heat by radiation and temperatures fall rapidly. In the cool season both the maximum and minimum temperatures are lower and night fires are frequently lit to provide warmth. The figures shown on the graphs for Cairo can be taken as fairly representative of this type of climate.

Natural Vegetation

Certain specialised forms of plant life can exist where a small amount of moisture is available in the deserts. Many plants are capable of storing water for long periods. These include the cactus, euphorbia and aloe. The *cactus* has spreading roots which collect any rainwater that may fall. The water is stored in the thickened stem whose tough skin restricts transpiration (the loss of moisture through the pores). There are many forms of cactus and most of them have spines in place of leaves, as a protection against animals. Various types of stunted bush, commonly referred to as *scrub*, also occur

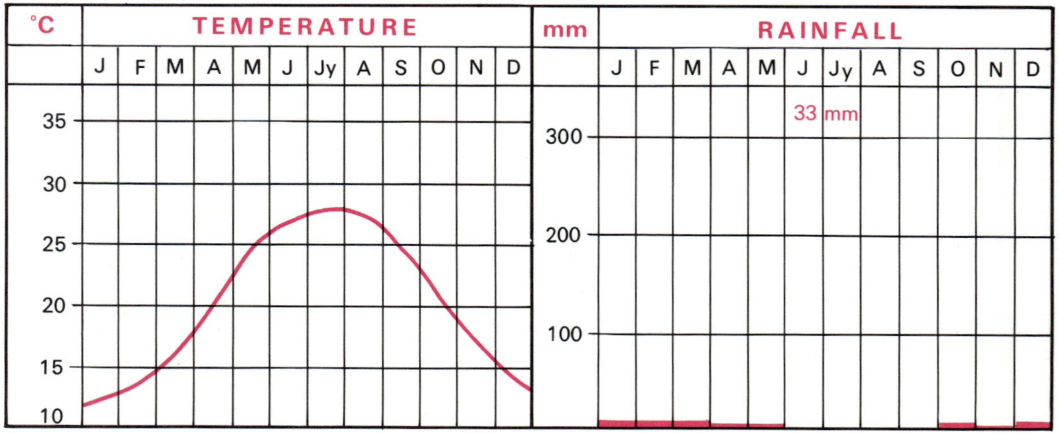

Figure 14.1 Climate graphs for Cairo

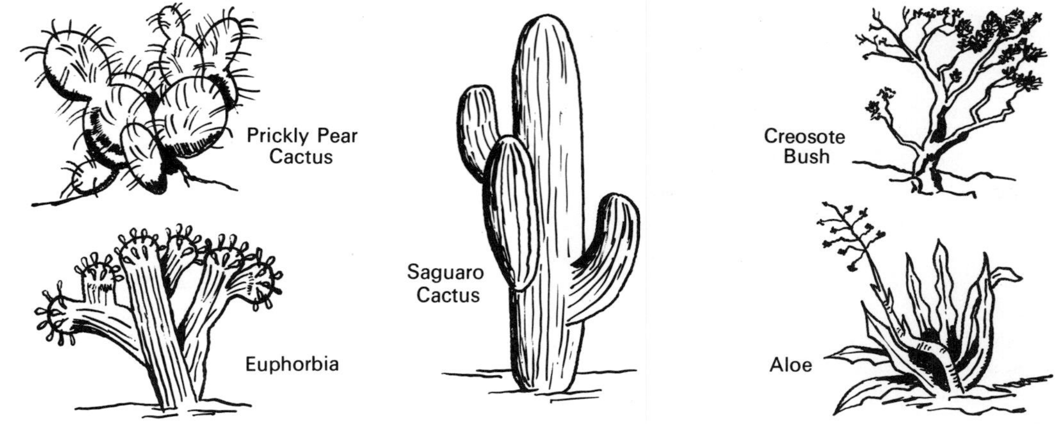

Figure 14.2 Desert vegetation

in the deserts. These use little water and grow far apart, though beneath the surface their long roots spread far and wide. Another fairly common sight is a carpet of grass following a shower of rain. The grass seeds germinate quickly and complete their life cycle within a few weeks, leaving fresh seeds to await the next shower.

Where water is present at or near the surface an *oasis* is found. The vegetation here would be rich and varied, with many types of palm, but most of the land is carefully cultivated, and the commonest palm is the valuable date palm.

The Saharan and Arabian Deserts

A vast desert stretches across North Africa from the Atlantic to the Red Sea, continuing across Arabia to the Persian Gulf and beyond. Its great extent is due to its situation in relation to the Asiatic land mass. In a way it can be said to lie in the rain shadow of southern Asia. Much of the surface consists of plateaus crossed by mountain ranges and containing numerous depressions. The inhabitants are mainly Arabs and the majority of them earn their living in one of two ways.

Firstly there are the *Bedouin nomads* who live in small family groups and are compelled to wander from place to place in search of pasture for their camels, sheep, goats and donkeys. Their possessions, apart from their animals, are few and include tents of wool or hair, blankets to keep them warm during the cool nights, goat-skin water bottles and a few simple domestic utensils. Their food is principally milk, meat and cheese obtained from the animals they herd, and dates and grain which they obtain by barter from the oasis dwellers.

Their most useful animal is the *camel* which can go without food and water for many days. It stores water within its body and fat in its hump, and its tough hard mouth allows it to eat thorny leaves and other desert plants. Its feet have broad pads to enable it to travel over sand, and during a sandstorm it can close its nostrils and protect its eyes by lowering an inner eyelid. Camels are used for both baggage and riding, and their milk, meat, hair and skins are all of great value to the nomads.

The second group of inhabitants are the sedentary (settled) peoples of the oases. They are frequently referred to as the *Fellahin* and live in permanent homes made of stone or dried mud. These are often crowded into a small space and surrounded by a wall which served in the past as a protection against bands of robbers. The flat roofs of the houses are often made by spreading palm leaves or straw over wooden beams. The larger houses are normally built round a courtyard.

A vital feature of life in an oasis is the carefully regulated use of water on the fields, known as *irrigation*. Many methods are used to draw water from wells and pools, including the two long-established methods shown in figure 14.3. The water is led along irrigation channels to the fields. A great variety of crops can be grown including wheat, maize, rice, sugar cane, cotton and beans. One of the most valuable trees is the *date palm*. It produces a highly nutritious fruit, and its fibrous trunk and broad leaves are used in building, thatching and for fuel. Another advantage is the shade it offers from the burning heat of the sun.

Egypt and the Nile

If it were not for the Nile, Egypt would be just

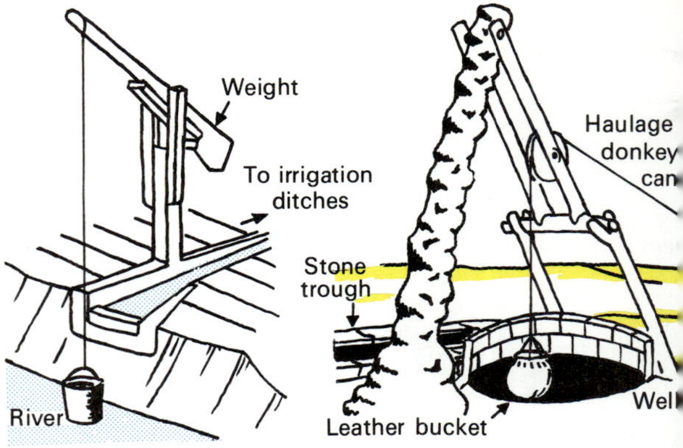

Figure 14.3 Low-technology methods of irrigation

top left Vegetation on the edge of the Atacama Desert in Chile at about latitude 30° south. *bottom left* Comment on the human activity, desert surface and buildings in this scene on the edge of the Nile-irrigated lands near Luxor, Egypt. (*Photos H R Singleton*)

another part of the desert. In fact about 30 million people are fed from the irrigated lands that border the river. The Nile is subject to great fluctuations in level. The main river, the White Nile, flows from Lake Victoria on the plateau of East Africa, so that it receives a good supply of water throughout the year, but it then flows through the Sudan, with its summer rainfall, before reaching Egypt. In addition, the summer rains of the Ethiopian Highlands increase the flow of two important tributaries, the Blue Nile and Atbara, which used to cause the annual floods along the lower Nile.

The traditional method of irrigation, which is often called basin irrigation, has been to hold back the flood waters in small embanked fields. The soil

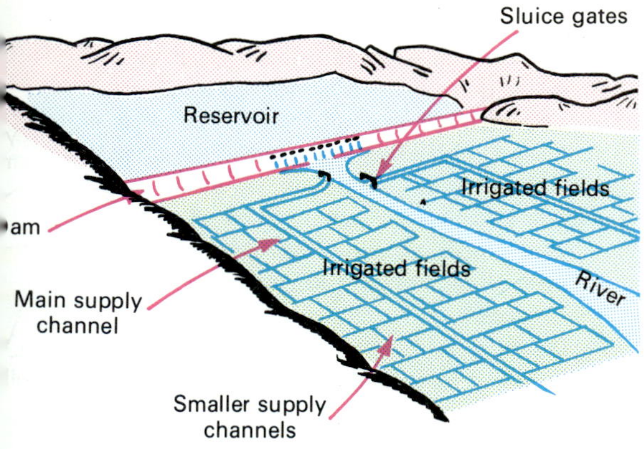

Figure 14.4 A modern perennial irrigation scheme

is enriched by the silt, but crops can only be grown in autumn after the floods have subsided. More important today, however, is perennial irrigation, which involves damming the Nile at several points including the great Aswan Dam. This holds back large quantities of water, ensuring a regular supply to a much larger area by means of a vast network of irrigation canals and ditches. Two crops per year and sometimes even three can be grown on the same land, the principal ones being cotton (which is of a high quality), maize, rice, sugar cane, millet, wheat and dates.

The first Aswan Dam was built by the British in 1902, but the newer Aswan High Dam has created Lake Nasser, which is one of the largest artificial lakes in the world. As a result of the new dam, Egypt's cultivated area has been increased by one quarter, the annual floods have been prevented, and the large hydro-electric station alongside the dam has doubled the country's electricity supply.

Other Desert Areas

Large-scale perennial irrigation is also practised in Mesopotamia (Iraq) where the waters of the Tigris and Euphrates are extensively used, and in Pakistan where an enormous area in Sind is irrigated by water impounded behind the Sukkur dam on the River Indus. In the United States valuable crops of sugar cane, cotton, fruit and cereals are produced on land watered from the Colorado and Gila rivers. Here on the Colorado is one of the highest dams in the world, the Hoover (Boulder) dam, whose maximum height is 221 metres.

Many of the hot desert regions possess very valuable *mineral deposits*, a feature which often attracts considerable populations to particular areas. Oil is probably the most important mineral asset of these regions, examples being the oilfields of Arabia, Iran and Iraq, and the newly developed ones in the Sahara. The Atacama desert in Chile provides nitrates (a valuable fertiliser), the Kalahari desert in Namibia diamonds, and the Australian desert gold, iron ore and many other minerals.

Irrigated land in Southern California
This was desert some years ago, but the area now produces a great variety of fruit and vegetables. Much of the citrus fruit, such as oranges, is now crushed and canned before distribution throughout the country (*U S Information Service*)

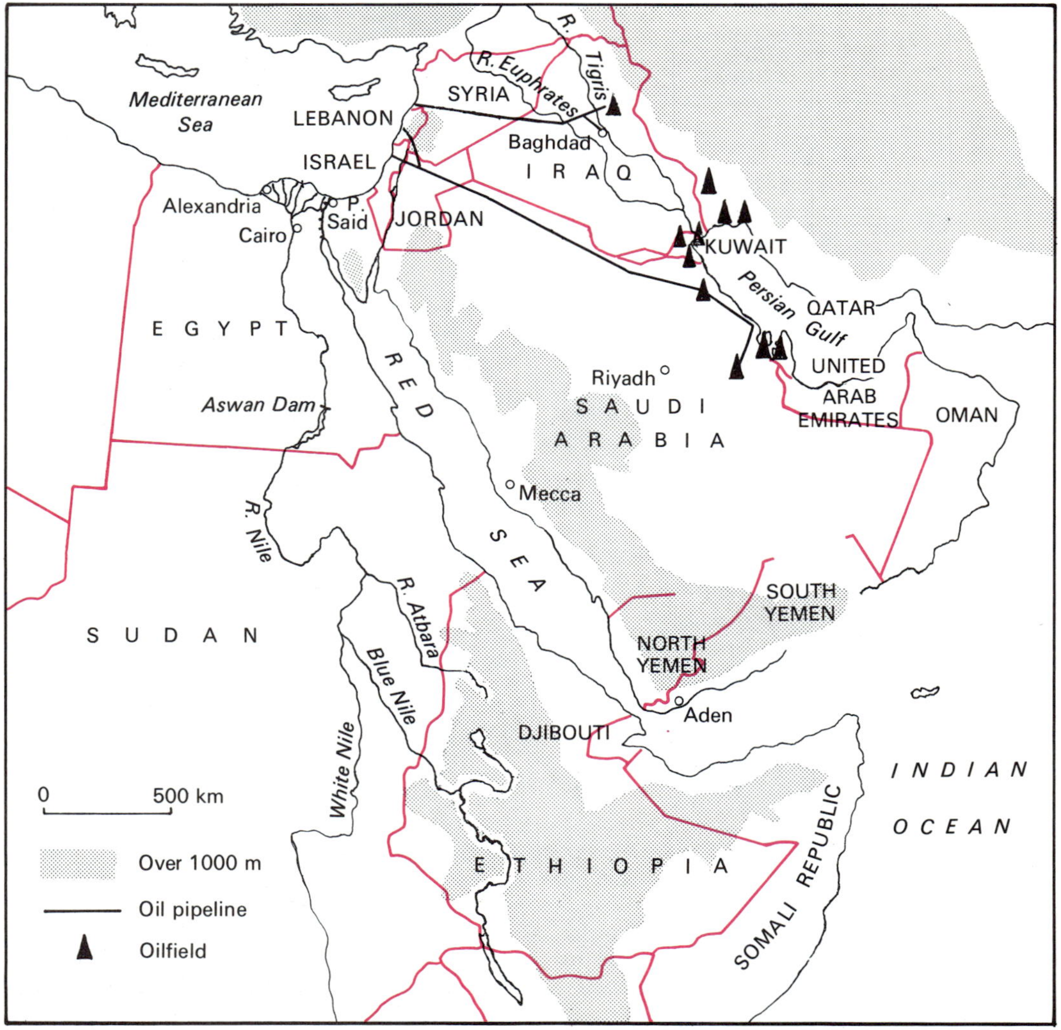

Figure 14.5 The Middle East

Exercises

1 Name and explain a major problem which occurs in irrigated areas in deserts and which is a result of high evaporation rates.

2 The Aswan High Dam has stopped the annual flood of the River Nile. As a result silt is no longer spread naturally over farmland.
(*a*) What problem results from this?
(*b*) How does this problem have to be overcome?

3 Explain why:
(*a*) There is a large daily temperature range in hot deserts.
(*b*) Rainfall is very low in hot deserts.

4 Make a plan of a typical North African oasis based on the following description: 'The oasis, about 2 km across, has developed round a number of wells which supply water for the numerous gardens and date palms. Nearby is the village, with many mud-built houses crowded within an outer wall. A taller building is the mosque. The dividing line between the oasis and desert is sharply defined. Just outside it is the village burial ground.'
Add a key and scale to your map.

Chapter 15

Monsoon Lands

Climate

The boundaries of the monsoon lands are shown in figure 15.2. It will be noticed that there are large areas both within and outside the Tropics. These are considered together in this chapter since they are under the same major climatic influences and have many other similarities.

In the hot season an intense low-pressure system develops over the interior of Asia causing warm, moist air from over the Indian and Pacific Oceans to move towards the land. This brings heavy rain in the middle of the hot season. The figures for Calcutta show the hot dry season as lasting from March to May. This is a period of cloudless skies and great heat, with mean temperatures rising to over 30°C. In June the rains break and temperatures are slightly lower. It may rain almost continuously for the first two weeks, after which there are heavy showers falling every few days. The wet season ends in October when the pressure distribution becomes reversed. A high-pressure system develops in the interior and dry winds blow out from the land. Temperatures between November and March are lower. In Calcutta the mean temperatures for both December and January are below 20°C.

Thus we can distinguish three seasons, a hot dry season from March to May, a hot wet season from June until October, and a cooler dry season from November to February. This is for tropical areas. In temperate monsoon areas the same seasons are recognisable but temperatures are lower and the winters can be severe.

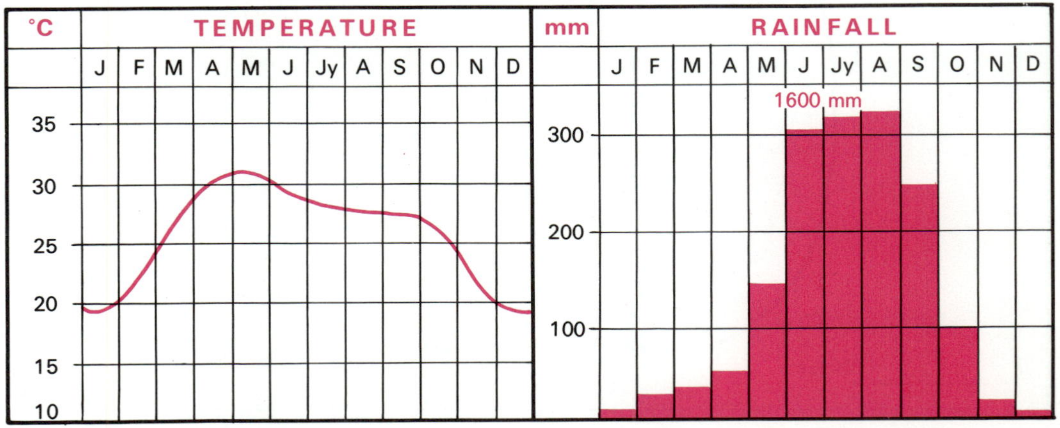

Figure 15.1 Climate graphs for Calcutta

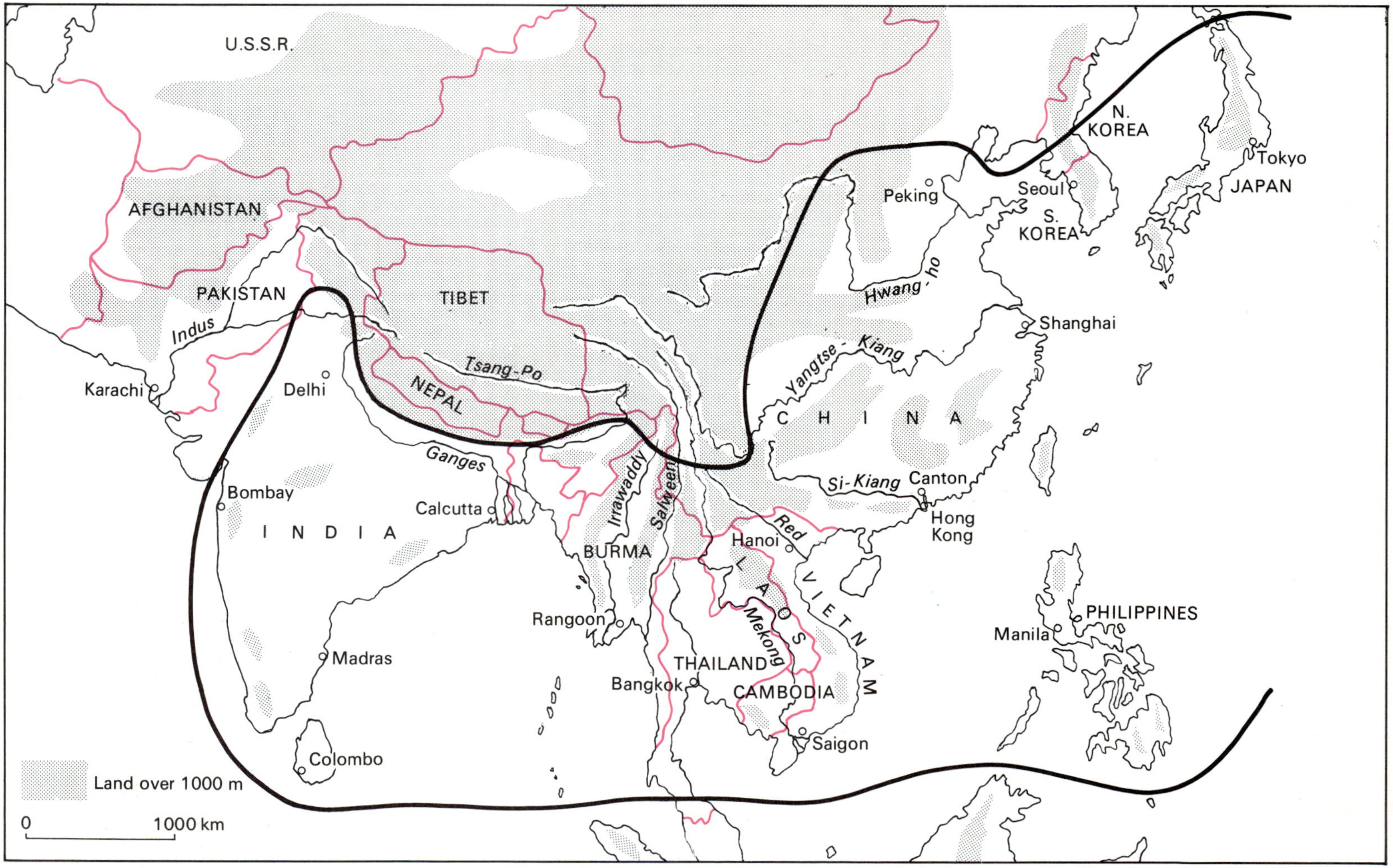

Figure 15.2 Monsoon lands of South-East Asia

The amount of rainfall varies greatly. Where rain-bearing winds meet high mountains, as they do along the stretch of the Himalayas north of the Bay of Bengal, the rainfall may exceed 5000 millimetres per annum. Other portions of the monsoon region, on the fringes of deserts, have rainfalls of below 500 millimetres.

Natural Vegetation

In tropical monsoon areas where the rainfall exceeds 2000 millimetres per annum evergreen forest similar to that of the equatorial region is found, but in areas experiencing between 1000 and 2000 millimetres there is a more open type of monsoon forest, mainly of deciduous trees which shed their leaves during the period of drought. One of the most valuable of these deciduous trees is teak, a hardwood in great demand on account of its water-resisting qualities. The drier areas, where the rainfall is less than 1000 millimetres, have a vegetation similar to that of the savanna, consisting of thorn bushes, scrub or tall grass.

Much of the natural vegetation has been cleared from the more fertile monsoon areas, particularly in the great alluvial basins which are among the most densely populated areas in the world. Deforestation of the remaining monsoon forest is taking place as quickly as in the tropical rain forest, with resulting problems of soil erosion in mountain areas, flooding in lowland areas and deposition of *coarse* sediments by those floods instead of fine alluvial sediment.

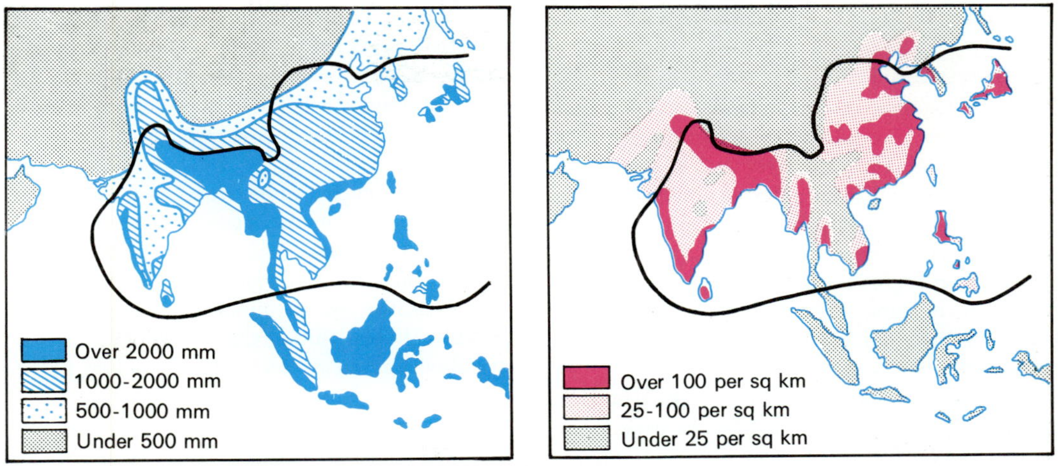

Figure 15.3 Annual rainfall and distribution of population

Figure 15.4 The monsoon winds of India

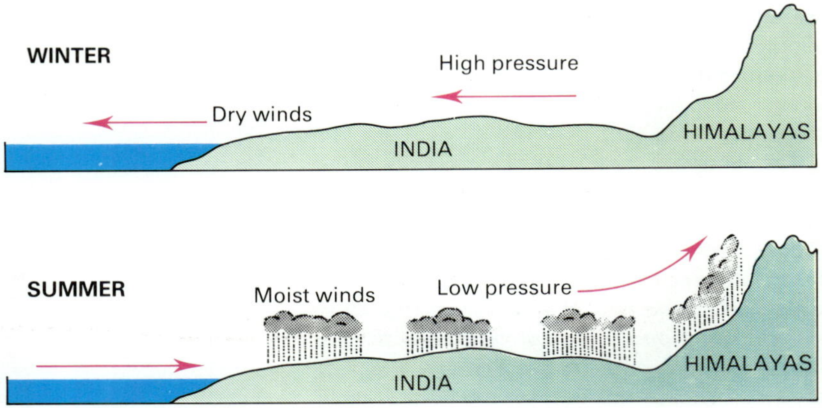

SOME TYPICAL MONSOON AREAS

The Ganges Basin

The River Ganges rises in the Himalayas and later flows for over 1600 kilometres across the northern plains of India. Here the river and its tributaries have spread a thick layer of fertile alluvium during the annual floods. Where the rainfall is heavy, in the lower Ganges and delta regions, the rapid growth of crops and a good yield are assured. This is true also in some of the drier areas of the upper Ganges where a reliable irrigation system has been developed.

Wherever possible the site of a village is on land above the flood level. The type of house depends on the locally available building materials. The poorer houses may be of bamboo plastered with mud, the roofs being of straw, reeds or palm leaves. The wealthier villagers have homes of brick with tiled roofs. These bricks are made from mud, baked in the sun. Communications from one village to another and to the surrounding farmland are by boat along the rivers and canals, or by bullock cart along roads which are dusty in the dry season and quagmires in the wet.

The farms are very small, the majority being under three hectares and the chief crops being rice, jute, maize, millet and sugar cane. *Rice* is the staple food of most of the people. Its yield is greater than for any other cereal, and it has been calculated that under half a hectare of rice can permanently support two people. For most varieties of rice the land must be flooded for the first few weeks of growth. To make this possible the farmers build low mud banks round the paddy fields to contain the water and prevent it from draining away. Silt deposited near the rivers helps to maintain the soil fertility, but rice rapidly exhausts the soil. All available compost is taken to the fields but in most cases yields could be greater with the more widespread use of fertilisers.

When the rains begin the rice seeds are sown in a nursery bed, and in the meantime the fields are prepared. The banks round each field must be checked and the field ploughed. The wooden plough, drawn by a bullock or water buffalo, churns up the soil, though the hooves of the animal have almost as much effect as the scratching of the plough. Then a flat board is dragged across in order to level the field. Women and children transplant the young rice plants. The crop is ready for harvesting in about four months. Cutting is done mainly by sickle which is a laborious process. The rice is carried to the courtyard where it is threshed

right **Rice planting, Madras State** (*Paul Popper Ltd*)
far right **Threshing** (*Ewing Galloway Ltd*)
Study the three photographs which illustrate various activities associated with rice cultivation. In what ways do the methods show that farming is intensive? What would be the effect of mechanisation on farming?

Harrowing (*Paul Popper Ltd*)

either by beating it with flails or trampling it beneath the hooves of animals. Finally it is thrown in the air to remove the chaff.

The system of farming is known as *subsistence farming*. It means that farmers and their families produce food for their own requirements but little surplus to sell. Some other crops including vegetables are grown, particularly where water is available from wells and pools to irrigate the land during the dry season. A crop of great importance in the Ganges delta region, near Calcutta, is *jute*. This is a tall marsh plant which provides a fibre used mainly in making sacking. In addition to these crops oxen and goats are kept grazing mainly on waste land.

Modernisation of farming can have a great effect. One development has been the introduction of high-yielding varieties of rice (HYV). In some areas rice yields have increased by over 150%. These new varieties do need quite large amounts of fertiliser and applications of pesticides to protect the plants from pests and diseases. In addition, the amount of water needs to be controlled carefully, so areas with irrigation systems for controlling water benefit most. One great advantage of HYV besides higher yields is its shorter growing season. This allows a second crop each year if water is available.

Areas relying on direct rainfall are unable to use HYV in the main growing season, as the water is too deep, but where water is available in the dry season HYV has become popular.

New varieties of rice are not the only forms of modernisation that have been introduced to raise living standards. Improvements in water supply, provision of reliable electricity supplies, mechanisation, especially of threshing where manual methods lead to wastage, better methods of storage, more widely available advice on new methods, and the availability of loans so that farmers can put new methods into practice are changes that have to go together. Increased production also means more food for sale, so that farms become more commercial and, with more intensive production, provide regular work for their many landless workers.

Tropical crops

1 **Sorghum** (*H Thomas*)
2 **Pineapple** (*S Arnold*)
3 **Cocoa beans** (*S Arnold*)
4 **Yam** (*S Arnold*)

Write a paragraph on each of the vegetable products illustrated, explaining the differences in preparation for consumption and the type of market each is destined for.

5, 6 **A tea plantation in the Kericho district of Kenya** (*Brooke Bond Liebig Ltd*)
Although India, Sri Lanka and China are noted for tea production, its growth is not confined to monsoon regions. The savanna areas of East Africa can provide very similar conditions.

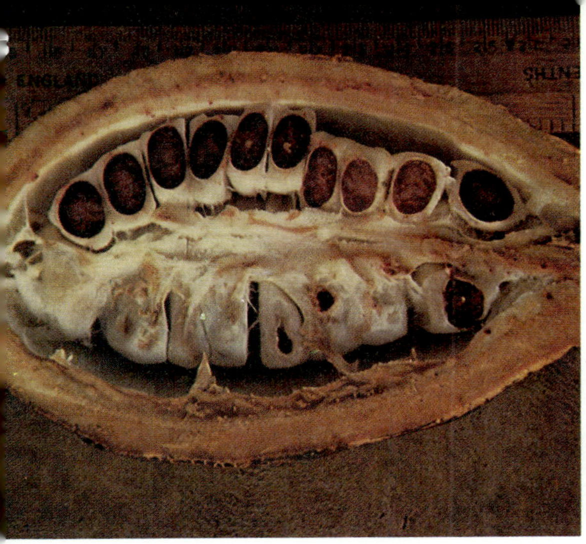

5

6

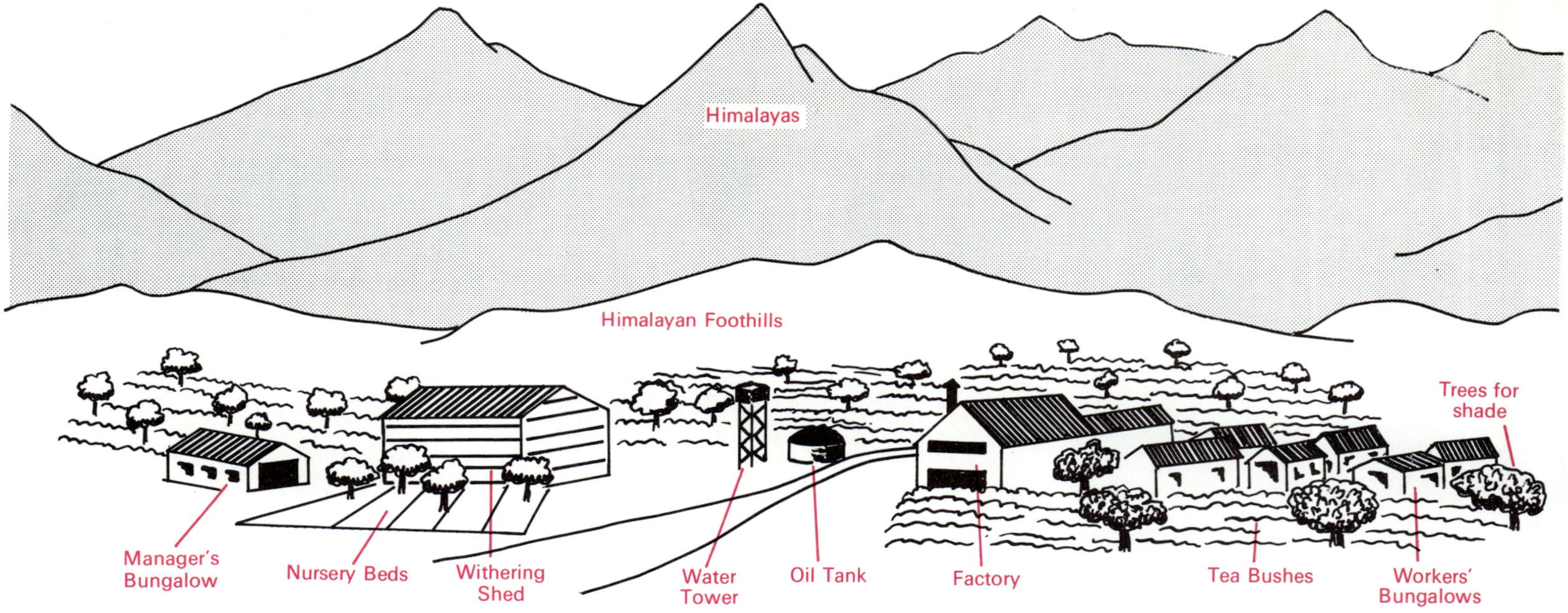

Figure 15.5 A tea estate near Darjeeling

Figure 15.6 The production of tea

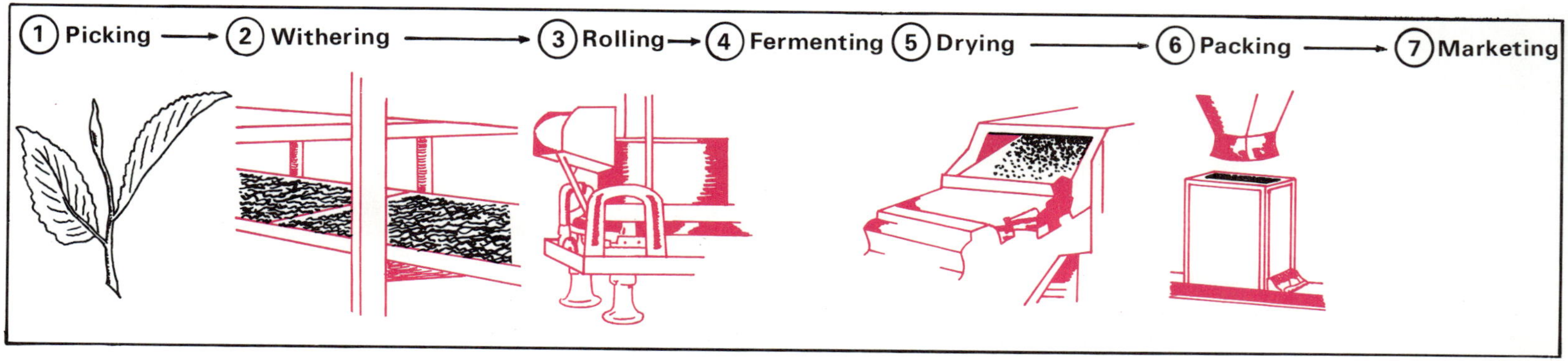

The Tea Gardens of the Himalayan Foothills

North of the Ganges plain along the foothills leading to the Himalayas are stretches of dense jungle. In some parts (for example round Darjeeling) these have been cleared for the cultivation of tea, which is grown on large plantations employing many workers. The south-facing slopes of the Himalayas have abundant sunshine and rain, and are well drained. These conditions are ideal for the tea bush. Work on the plantations is carefully organised. Men are employed to clear the land, break up the soil and look after the nursery of young cuttings. Once the young bushes have established themselves they are pruned back so that the new shoots will multiply and can easily be picked. Most of the picking is done by women, and the young leaves are taken to the plantation factory where they are passed through various processes including withering, rolling, fermentation, drying, sorting and packing in tea chests. The latter are then distributed to many parts of the world. Tea is also exported from southern India and Sri Lanka. There is a large production in south China, but only small quantities are exported.

The Drier Areas of India and Pakistan

Travelling west from the Ganges basin one enters the Indus plains. This is a hot dry region where irrigation is vital for most crops. Water is obtained by leading off irrigation channels from the Indus and its tributaries. Most water is obtained in April and May when the rivers are swollen by the melting snows of the Himalayas. Here wheat is the most important crop. In the past there has often been famine, and to overcome this many big new irrigation schemes are being developed.

South of the Indo-Gangetic plains the Indian peninsula is occupied by the plateau of the Deccan. Apart from the well-watered coastal plains between the Western Ghats and the Arabian Sea much of the Deccan is relatively dry. The north-west part has a fertile black soil, of volcanic origin, which is noted for cotton. Farther south, between the Western Ghats and the Bay of Bengal the tank method of irrigation is used. This involves the construction of earthen banks across small valleys so that lakes are formed when the rains come.

Throughout India oil seeds are important. These include linseed, sesamum and, in the drier areas, groundnuts.

Other Tropical Monsoon Areas

The alluvial lowlands of Burma, Thailand, Vietnam and south China have much in common with the Ganges plain. Their populations practise an intensive form of agriculture based on the cultivation of rice. In parts there is dense jungle, especially along the hill slopes.

Northern Australia also has a tropical monsoon climate, although the area of heavy rainfall does not extend far inland. Undoubtedly the region could support a considerable population. Rice growing has been tried, without great success, and there are large cattle stations on the interior plateaus.

Temperate Monsoon Areas

The temperate monsoon regions of north and central China, Korea and Japan have a surprisingly similar appearance to the tropical monsoon areas in spite of the cold winters. This is the first natural region we have dealt with so far where the word 'winter' is more apt than 'cool season'. The lowlands here are again densely populated and produce wheat, rice and maize. In north China the soya bean is a staple crop.

Japan is so densely populated, and its lowlands so restricted, that manufacturing industries have been developed since the beginning of this century. Today Japan is one of the world's leading industrial nations whilst both China and India are making big efforts to follow this lead.

Exercises

1 How does the rainfall pattern in a monsoon region differ from that in a savanna region?

2 Agricultural production has increased greatly in some parts of the Ganges Basin. In part, this is due to the modernisation of farming.
   (*a*) What are the different ways in which farming has been modernised?
   (*b*) Choose *one* of the features you mentioned in answer to (*a*) and explain fully how it has helped in raising living standards.

Chapter 16

Warm Temperate Western Margins

(Mediterranean Regions)

Climate

The warm temperate western margins lie on the western sides of land masses between latitudes 30° and 40° north and south. We have already mentioned in Chapter 4 the seasonal movement of the world pressure and wind belts. This causes the Westerlies to blow from the sea over these regions in winter and the Horse Latitudes high-pressure belt to dominate them in summer. For this reason the winters are mild and wet and the summers hot and dry, with abundant sunshine.

The figures for Palermo, Sicily, show mean monthly temperatures in the winter of between 10°C and 15°C, and in the summer between 20°C and 25°C. Of the total rainfall of 757 millimetres 80% falls between October and April.

Natural Vegetation

Shallow-rooted plants including grasses are not favoured by the climate. Trees and shrubs are mainly evergreens and have adapted themselves against the summer drought in a number of ways. Most have very long roots which reach down to the subsoil in search of water. The leaves are small, leathery and often wax covered so that transpiration is restricted, and the bark is often thick. The cork oak, which has an outer cork bark over $2\frac{1}{2}$ cm thick, the wild olive and the small Mediterranean pine are typical trees.

Large areas are covered with a dense mass of low shrubs including laurel and myrtle, and are known as *maquis* in southern France and Corsica. Limestone areas with a thin soil are known as *garrigue* and have a scrub vegetation of broom, gorse and aromatic plants like lavender and thyme, with much bare rock between.

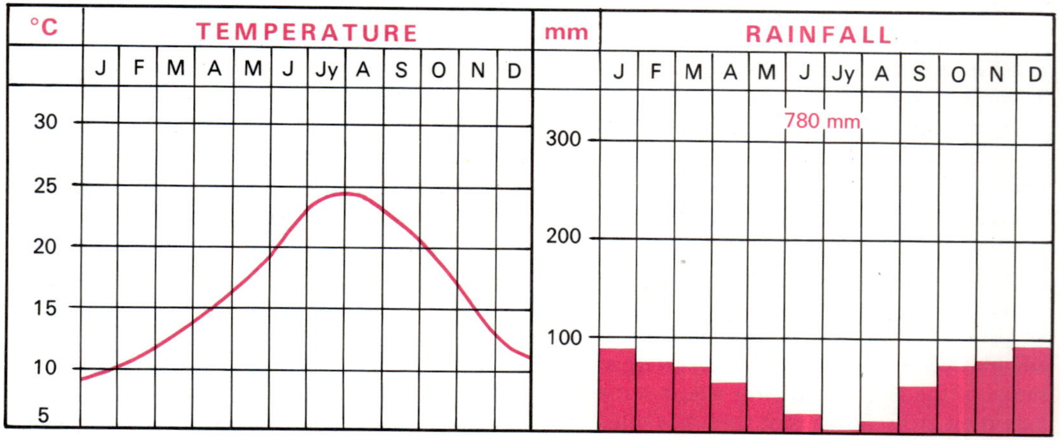

Figure 16.1 Climate graphs for Palermo

Mediterranean vegetation (*F R Dobson*)

Irrigation in Mediterranean lands (*F R Dobson*)
In what ways is the natural vegetation adapted to a Mediterranean climate? Describe the scene above. What crops are likely to be grown under irrigation? Why is irrigation necessary?

THE MEDITERRANEAN SEA AREA

The generally favourable geographical conditions have played a large part in the early development of civilisation in the countries bordering the Mediterranean Sea. Today many of these lands are well developed agricultural areas with a high density of rural population and noted for the production of fruit and vegetables. Although there are some regional differences the methods of farming are everywhere intensive, i.e. aimed at obtaining the maximum yield per hectare. There is little mechanisation and the fields are normally small, the land on steep slopes being frequently terraced. Irrigation is practised wherever possible during the dry summer. The water is obtained from streams which do not dry up or from wells.

Cereals, particularly wheat, are important for local consumption. Wheat is planted in the autumn so that it can make growth during the moist winter and spring and is then ready for harvesting early in the summer. It is of a hard variety and is often eaten in the form of spaghetti. Maize and barley are spring planted for later harvesting and there is an increasing production of rice on the larger alluvial plains.

Rows of olives and grape vines are a typical sight in all these lands. Sometimes the ground in between the rows is cultivated for pumpkins, melons, tomatoes, lettuce, beans and other vegetables.

The *grape vine* requires a great deal of attention throughout the year. It must be manured and pruned and, for many varieties, has to be trained on wire or wooden supports or, as often happens, over the olive trees. Then it must be sprayed as a protection against pests and diseases. Harvesting takes place in September, and farmers take their grapes to the local centre to be crushed ready for

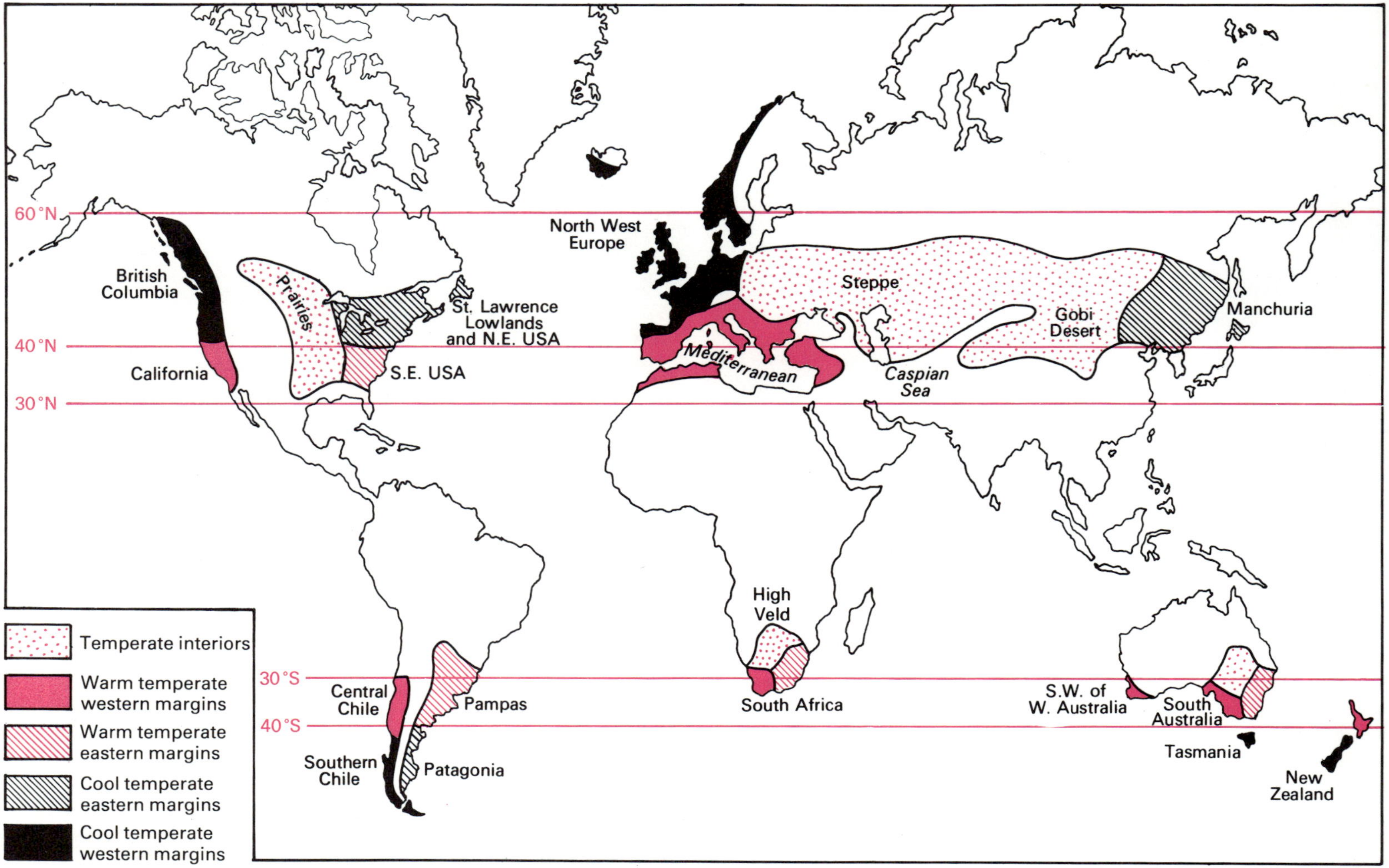

Figure 16.2 Temperate regions

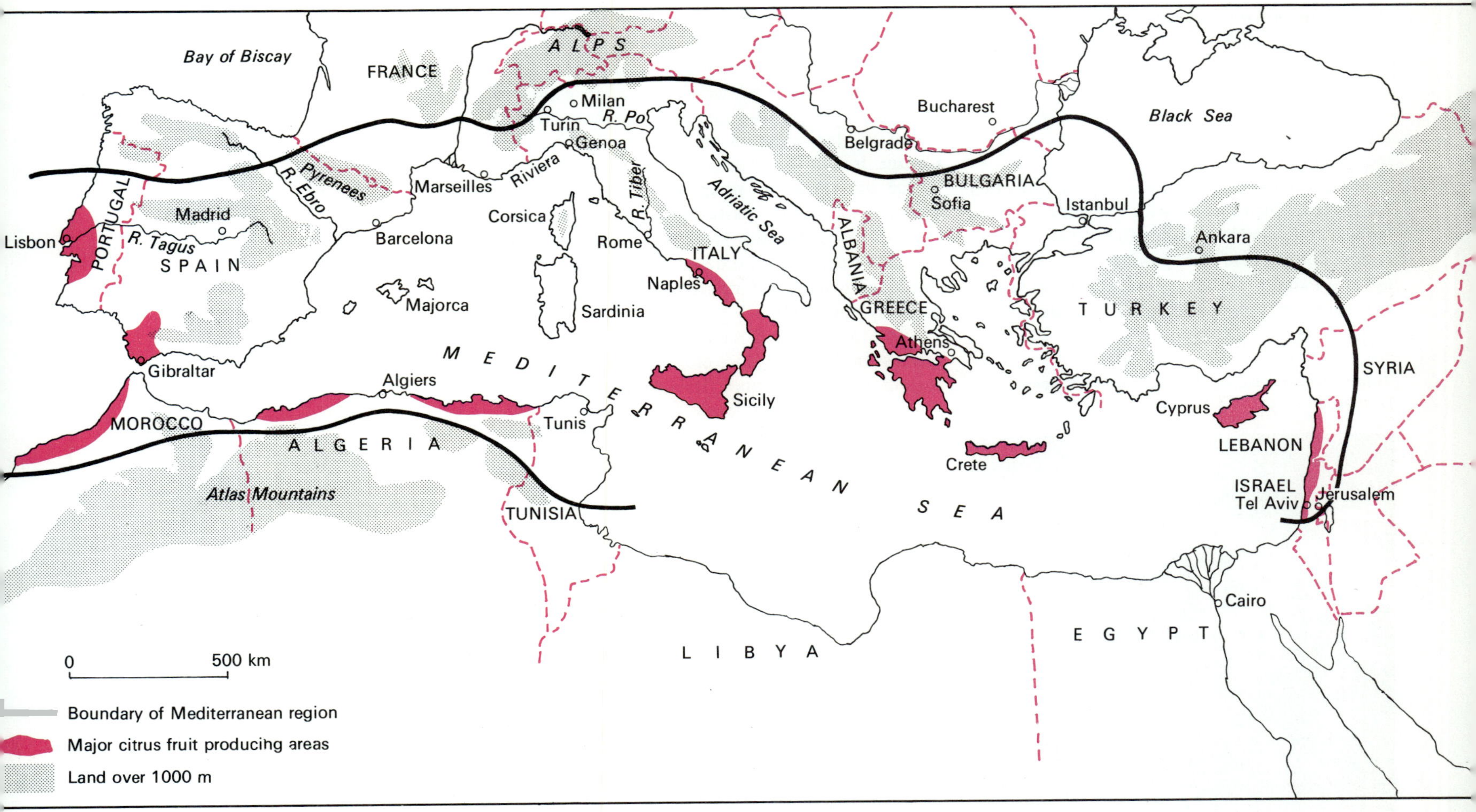

Figure 16.3 The Mediterranean Sea

the making of wine. Some grapes are used as fresh fruit and others are dried for currants and raisins.

The *olive tree* is widely grown, especially in Spain, Italy and Greece. It requires little water and is grown in groves, often on hillsides. Both the oil and the fruit itself provide an important part of the local diet.

Besides olives there is a great variety of other fruit trees. Citrus fruits which include oranges, lemons and grapefruit, although not native to this region, flourish here as do peaches, pears, apricots, figs and walnuts. These may grow in groves or, like the vines, in well-spaced rows providing some protection from the sun for the vegetables grown in between. The lowlands of south and south-east Spain have produced oranges for many centuries under a carefully regulated system of irrigation. Orange-growing is also important at the eastern end of the Mediterranean, particularly in Israel and Cyprus. Italy is the principal producer of lemons.

Tourist development on the Costa del Sol, Spain
(*D Jones*)

The movement of sheep and goats between summer and winter pastures, or *transhumance*, used to be widespread. However, it has been in decline for many years and is now limited in extent due to more intensive farming and the enclosure of formerly open land.

The Mediterranean region has seen a massive increase in the importance of *tourism*. The coastal regions of Spain, France, Italy, Yugoslavia and Greece now receive millions of tourists every year, mostly from other parts of Europe. In some areas development has been so extensive that coasts have become continuously built-up, like the Costa del Sol in southern Spain which extends almost without a break for over 50 kilometres.

OTHER AREAS

The other warm temperate western margin areas are well scattered and have been largely developed by European settlers, the British in Australia and New Zealand, the Dutch and British in South Africa, and the Spanish in Chile and California (before it became part of the United States).

California

For many years this was ranching country and later became a cereal-producing area. Today the advantages of the climate are being fully utilised, and

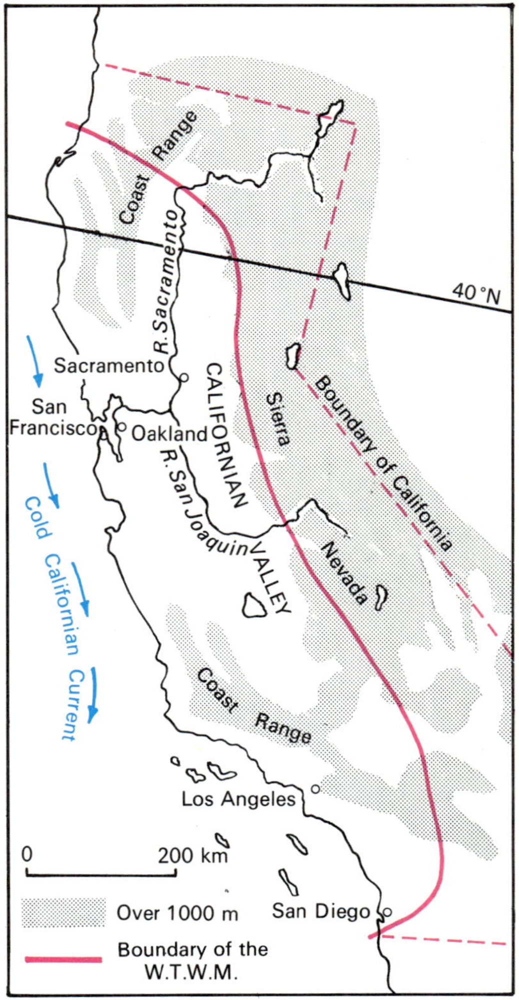

Figure 16.4 California

the southern Californian coastal lowlands and the Californian valley together make up the greatest single area in the world for the production of fruit and vegetables. Irrigation is an important feature, particularly in the southern part of the state. The water is obtained from the numerous streams flowing from the Sierra Nevada and Coast Ranges. Fruit is dispatched fresh or in dried or canned form to all parts of the United States and abroad.

Southern Australia and South Africa

Here farming is less specialised. Wheat is important, and a wide range of fruits is produced including grapes, apples, pears, plums, apricots, peaches, oranges and lemons. The canning industry has grown rapidly, and large quantities of tinned fruit and jams are sent to many parts of the world including the British Isles. The cultivation of the vine is also expanding, and wines from both these regions are now competing successfully in world markets with European wines. Fresh fruit from these areas reaches Britain when fruit from the northern hemisphere is out of season.

New Zealand

North Island, New Zealand, is generally included as a warm temperate western margin region although it has considerable rainfall in summer as well as winter. Consequently there is much good pasture in the coastal lowlands and cattle are kept both for meat and dairy products. The very mild winters enable the cattle to graze out of doors throughout the year.

Central Chile

This is one of the most densely populated parts of South America. It has an economy very similar to that of the European Mediterranean areas. Wheat, beans, onions and potatoes are grown for local use. Oranges, lemons, olives and grapes are grown, and some of them exported. Oxen are used as draught animals, and sheep and goats are herded in the foothills of the Andes.

Figure 16.5 Temperate regions of Australasia

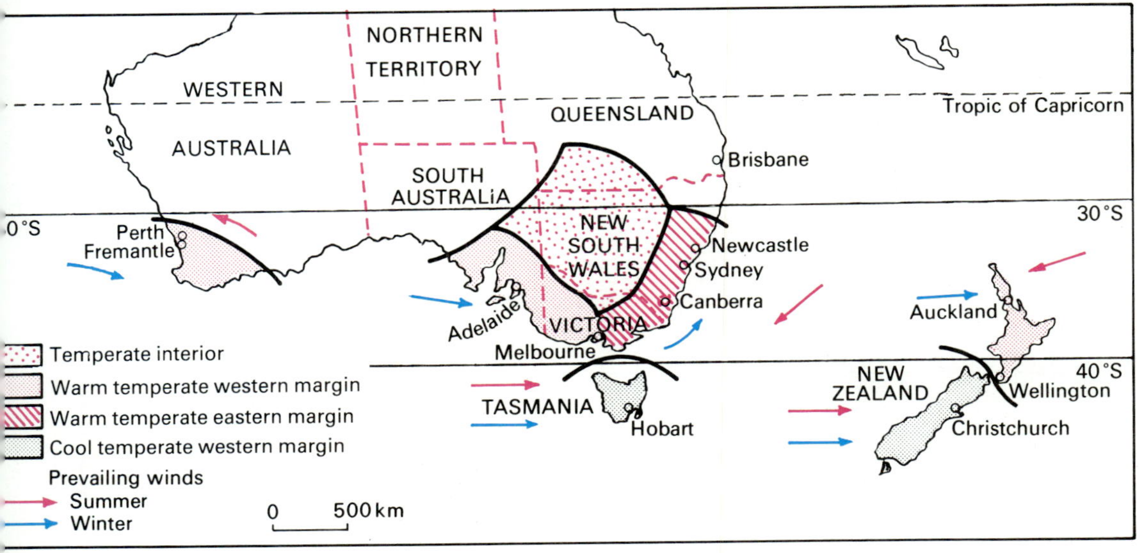

Exercises

1 Study figure 16.4.
   (*a*) Explain how the southern part of the Californian valley is provided with a reliable source of irrigation water.
   (*b*) Explain why irrigated agriculture in the Californian Valley is *intensive*.

2 The Mediterranean coasts of Europe attract millions of tourists each year.
   (*a*) Where do most of the tourists come from?
   (*b*) What are the attractions of these areas?

3 Study figure 16.3 and other maps showing industrial development in the same region.
   (*a*) Suggest reasons why the Mediterranean Sea in general and some parts in particular have become highly polluted.
   (*b*) Why is the pollution problem far greater for the Mediterranean Sea than it would be for coastal areas facing open ocean?

Chapter 17

Warm Temperate Eastern Margins

Climate

Lying in the same latitudes as the warm temperate western margins (30° to 40° north and south) these eastern margins include the south-east United States, the pampas region of Argentina and Uruguay, and parts of south-east Africa and south-east Australia. Temperatures are very similar to those of the western margins but the rainfall is much more evenly distributed throughout the year, with a summer maximum.

In winter westerly winds blow from the land and are therefore generally dry, although a little rain is caused by cyclonic disturbances. In summer winds tend to blow from the sea towards the low-pressure systems over the interior of the continents and bring a plentiful supply of rain. The bring a plentiful supply of rain. The climate graphs for Raleigh, in the south-eastern United States, are typical of this region except that the winter temperatures are rather lower than is general.

Natural Vegetation

The plentiful rainfall, especially in summer, has produced a natural woodland vegetation, consisting both of evergreen and deciduous trees. These include laurel, cypress, evergreen oak, chestnut and hickory, the latter being restricted mainly to the United States. The pampas region of South America is an exception, for here there are extensive natural grasslands, with woodland only in patches and restricted to the wetter parts.

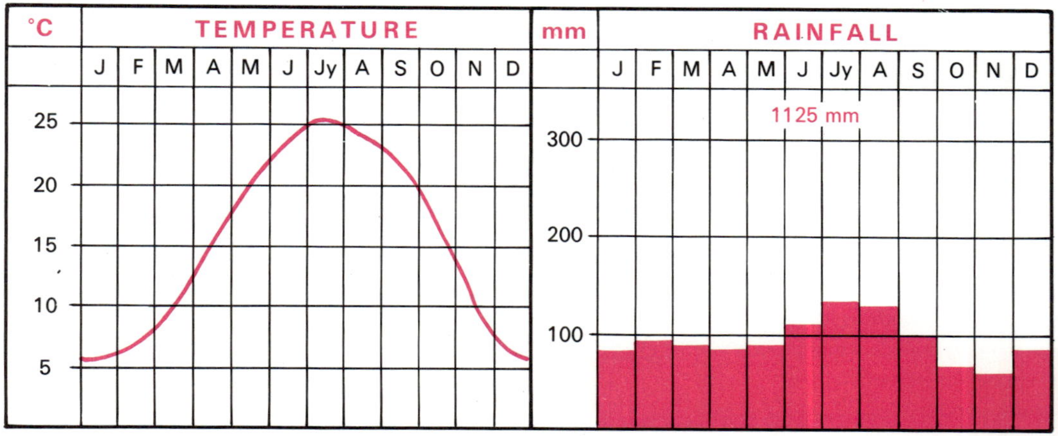

Figure 17.1 Climate graphs for Raleigh

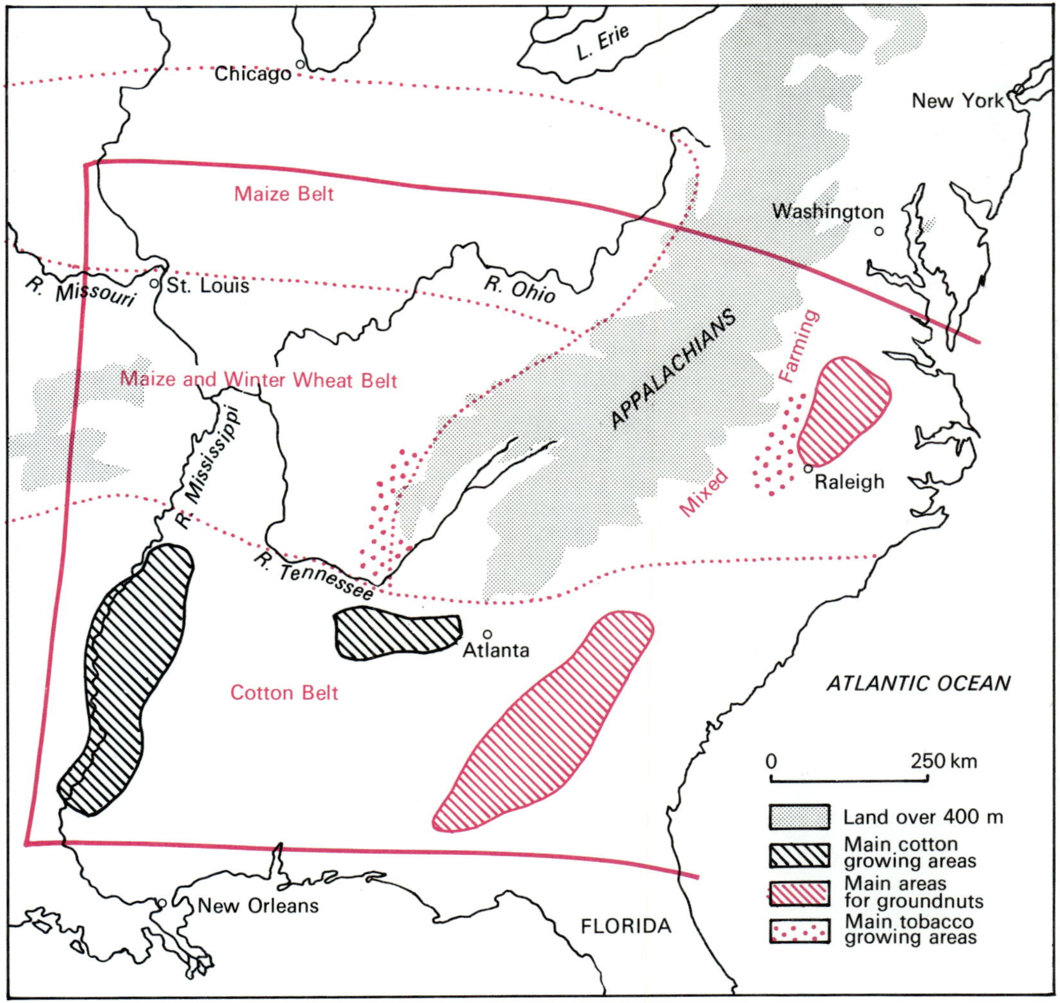

Figure 17.2 The warm temperate eastern margin in the USA

THE SOUTH-EASTERN UNITED STATES

This region is an important agricultural area. Part of it, in the South, is known as the Cotton Belt. In the past, cotton was the single most important crop and was grown on large estates as well as small farms. At various times cotton has been grown almost everywhere where the climate gave an assured frost-free period of 200 days, including areas of Texas where irrigation was used. Although the region still produces a very large proportion of the world's raw cotton, the crop is no longer totally dominant. Even the methods of farming have changed. Cotton-growing used to depend on large amounts of labour, but mechanisation was highly successful and cut labour needs drastically. The area of land used for cotton has fallen, but yields (the amount grown per hectare) have increased.

In the past the cotton-growing industry faced many problems, some of which were environmental and some economic. The major environmental problem was the boll weevil, a small beetle which devastated the crop in many areas. Spraying with insecticide is vital. The major economic problem has been changes in the price of cotton. This affected small farmers particularly badly, as in good years a glut of the crop brought prices and their incomes down.

The Cotton Belt no longer exists. Instead there are limited areas where cotton is the speciality crop.

There have been many changes in the agriculture of this southern area. The most widely grown crop now is the soya bean which is found in almost every part of the region. It is used in the food industry, as animal fodder in cattle cake, as a source of vegetable protein, together with many industrial uses. It is especially important in the Lower Mississippi Valley.

The changes have been more wide-ranging than simply involving the introduction of new crops. There has been a great increase in pasture land and the growing of fodder crops for beef cattle. Although dairying is carried on, it is beef cattle rearing that is most important and that has grown enormously, partly as a result of improved grasses and cross-breeding of British and Indian cattle to give greater production. There is now much more rotation of crops to meet the need to maintain soil fertility. Where cotton is still grown, for example, it is as part of a three-year rotation with hay and legumes or cereals. Poultry farming is very important and the region produces over half of the USA's poultry. This activity is highly specialised and is done on a large scale. Another major change is the great increase in the amount of land now under forest. Excellent stands of timber occupy large areas of land which were formerly used for crops.

Tobacco is a traditional crop that is still important. Virginia, Kentucky and parts of the Carolinas are major producers. The crop requires a warm climate and rainfall throughout the year. Other areas specialise in peanuts, peaches and rice.

Throughout the region *maize* is the most important cereal, especially towards the north-west (the maize belt). The crop is well suited to warm temperate regions which have an annual rainfall of 1000 millimetres or more. Here it can be assured of the minimum of 140 frost-free days required for its successful cultivation. Maize is actually the third-ranking world grain crop and is the leading grain

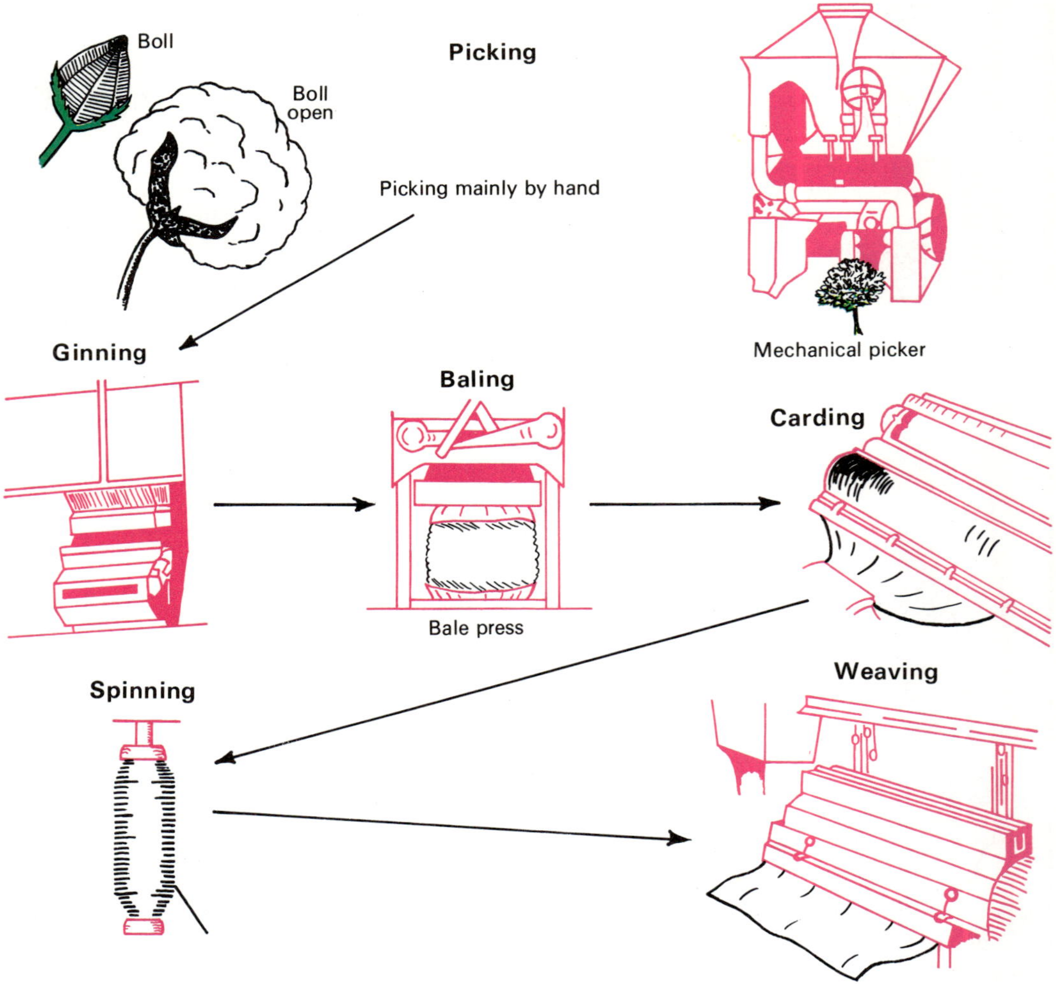

Figure 17.3 The cotton industry

The harvesting of maize (*U S Information Service*)
This scene is taken in the heart of the 'maize belt' in the USA. Varieties of maize are grown in many parts of the world. Where else can you think of, and in what ways does cultivation there differ from that in the USA?

fed to animals. Most of the maize produced is used on the farm for the fattening of cattle and pigs. This farming system is the result of high meat consumption in the United States. A typical maize, beef cattle and pig farm grows maize on about a third of the land, while on the rest oats, winter wheat, hay, vegetables and soya beans are produced. Cattle are brought from the high plains (which lie to the west of the maize belt) for fattening along with the pigs which are reared normally on the farm.

Although there are variations in the farming system from place to place, maize is a common element. The maize belt benefits from ease of access to markets, a high level of mechanisation, the lack of relief which in turn reduces soil erosion problems, fertile glacial soils as well as the highly suitable climate.

THE PAMPAS AREA OF SOUTH AMERICA

The pampas consists of the extensive alluvial lowlands stretching away from the shores of the Plate river in Uruguay and Argentina. The natural grassland has been improved to provide good grazing for beef cattle. The early Spanish breeds of cattle were replaced by British breeds to provide good quality meat for European markets. Railways helped to open up the area, and refrigeration led to the growth of enormous exports of beef.

Many of the farms consist of large estates called *estancias*, and the landscape has rectangular fenced fields with occasional water pumps to provide from wells and bore holes. As well as rearing beef cattle, the pampas region is an important producer of wheat and maize. The amount of land devoted to cereals or to beef cattle depends on world prices. For example, if cereal prices fall, more land is turned over to beef cattle. New crops have become important recently, in particular soya beans.

Within the pampas there are areas of specialisation. Dairy regions are found in several places, and potatoes, fruit and vegetables are grown to supply the big urban markets.

SOUTH AFRICA AND SOUTH-EAST AUSTRALIA

These smaller regions in South Africa, and New South Wales in Australia, are important for maize, fruit and cattle. In the coastal districts of New South Wales dairy farming is well developed to serve the large urban population of Sydney and the industrial region round Newcastle.

Figure 17.4 The warm temperate eastern and western margins of South America

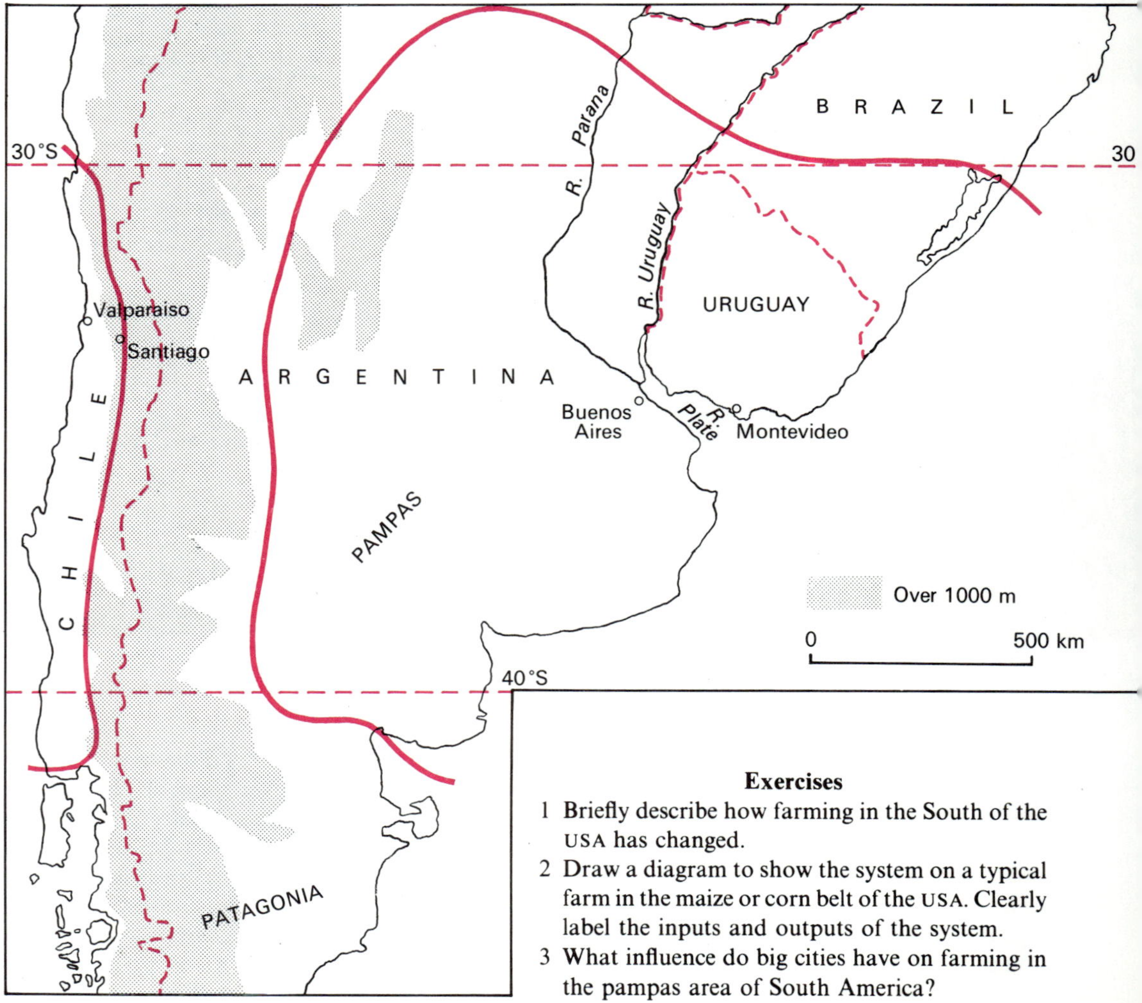

Exercises

1 Briefly describe how farming in the South of the USA has changed.

2 Draw a diagram to show the system on a typical farm in the maize or corn belt of the USA. Clearly label the inputs and outputs of the system.

3 What influence do big cities have on farming in the pampas area of South America?

Chapter 18

Temperate Interiors

Climate

The North American and Eurasian land masses are so vast that their interiors are a long way from the oceans which, as we have seen in Chapter 4, act as a moderating influence on the climate. Consequently the temperate interior regions in the northern hemisphere are noted for their great extremes of temperature.

This does not apply to the southern hemisphere where they are not so far from the sea and also lie nearer to the Equator, so that the climate is much less extreme.

In summer the great heat of the interior causes an ascending air movement which produces a low-pressure system so that winds are inblowing and bring light showers from the distant oceans. In winter the land radiates heat rapidly and a high-pressure system develops as a mass of cold air builds up over these regions. Precipitation at this time of the year is light and falls mainly as snow. The graph for Saratov in European USSR is typical. Temperatures, which average 20°C in summer, drop to well below freezing point during the five winter months. Saratov has an average annual precipitation of 378 millimetres, though in the region as a whole it varies from below 250 to about 600 millimetres.

Natural Vegetation

The rainfall is insufficient to support a forest vegetation so that grass is dominant. It grows rapidly in spring and early summer, developing long roots to combat the drought, but it dies down in late summer. Where the rainfall is below 250 millimetres the natural vegetation deteriorates to tufts of coarse grass and in places to bare rock and sand.

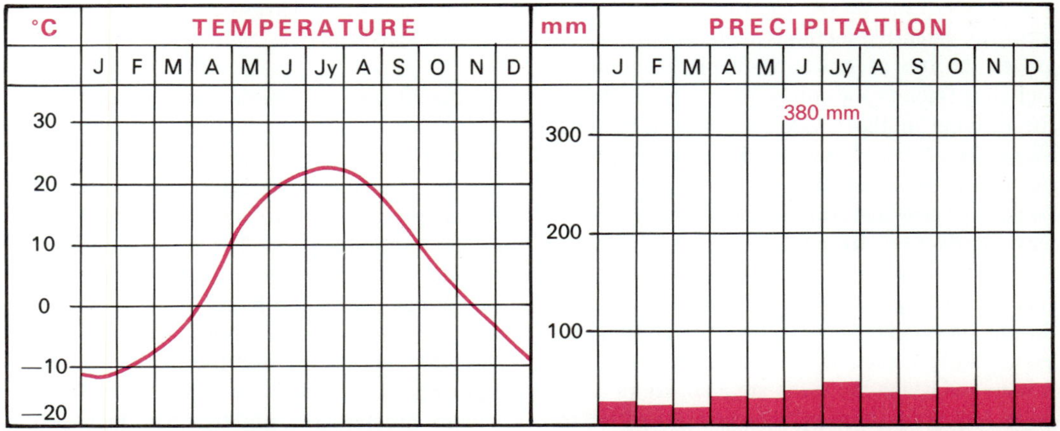

Figure 18.1 Climate graphs for Saratov

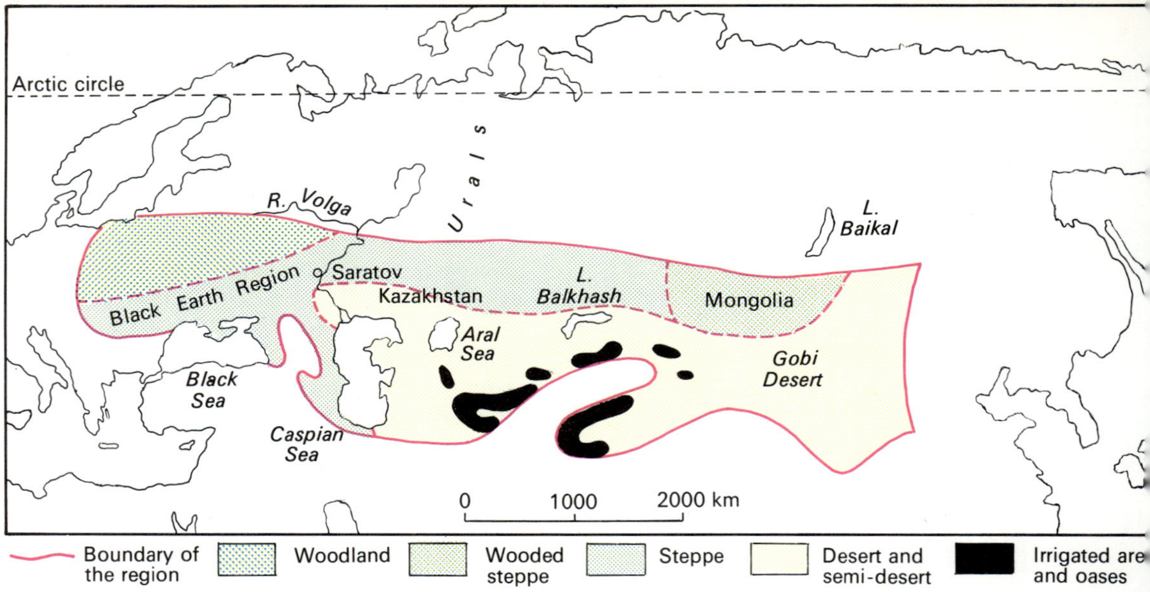

Figure 18.2 The temperate interior region of Eurasia

THE STEPPES

The steppes occupy an immense plain stretching from the lower regions of the Danube, in Hungary and Romania, across the Ukraine and into south-west Siberia, a distance of nearly 5000 kilometres. It is convenient to divide it into the richer black earth lands in the west and the drier steppe lands of the east.

The Black Earth Region

This includes the plains of Hungary, Romania and the Ukraine. Most of it has a rainfall of between 375 and 500 millimetres per annum, so that there are rich natural grasslands which have provided the soil with an abundance of humus (hence its black or dark brown appearance). This highly fertile soil is called *chernozem*.

The black earths of the Ukraine are the great grainlands of the USSR. The vast open plains are ideal for extensive methods of farming. This involves the use of much mechanised equipment including tractors and combine harvesters. Most of the farming in the USSR is in very large units known as *collectives*, in which the profits are shared out among the members, or in even larger state farms. The principal crop is *wheat* which requires at least 110 days free from frost in which to complete its growth cycle. The melted snow provides moisture for early growth, and the bright sunny conditions of summer are excellent for ripening the grain. Barley, oats, maize, sugar beet and sunflowers (of which the seeds are a source of edible oil) are also important in different areas within the region. The majority of farms have a mixed system, but grain-growing and livestock-raising are the two main elements.

The Dry Steppes of Siberia and Kazakhstan

Beyond the lower Volga the rainfall diminishes to less than 250 millimetres per annum, and at the same time the grasslands deteriorate and farming becomes precarious. Two major problems here are drought and soil erosion. The Soviet Government is trying to increase the arable hectarage in the lower Volga region and Kazakhstan. Most of the farms are very large *state farms*. They are usually larger than collectives, and the workers are state employees. State farms were set up to develop new methods and new land, and they have increased in number in recent years.

New varieties of grain have been developed which will survive in a dry climate. Tree belts have been planted as a protection against wind erosion, and the irrigated area is being greatly increased. Much of the region still remains primarily a sheep-rearing area.

In the northern regions of the west Siberian steppe, grass is more plentiful, and both grain production and cattle-rearing have increased. There is a growing market for food and raw materials in the new industrial towns which have grown up near the rich mineral deposits of the Urals and elsewhere.

The Semi-Deserts and Deserts of Central Asia

The poor grasslands deteriorate into semi-desert

east of the Caspian Sea. Here some of the people, such as the *Kirghiz*, are nomadic, although most have been settled by the government. Sheep, goats, horses and camels are kept, but cattle are now more important than in the past. They are driven to higher areas in the summer to graze when lower pastures are parched.

Many of the great mountain ranges of Central Asia rise above the snow line and provide water for irrigation schemes. In the irrigated areas a great variety of crops can be grown, and the Soviet Government has paid particular attention to cotton-growing in the Tashkent region. Heavy fertiliser applications are needed.

THE PRAIRIES

The prairies extend from the Great Lakes to the Rocky Mountains in both Canada and the United States. The natural conditions here are very similar to those of the steppes. In North America the rainfall decreases towards the west so that in the east it is sufficient to promote the growth of rich grass whilst in the west the grass is short and coarser. The prairies are undulating plains rising westwards from about 200 to 1000 metres above sea level. For a long time they were the hunting grounds for wandering tribes of Indians whose chief prey were the vast herds of bison. In the nineteenth century as the pioneers moved westwards the bison were killed off. At first large open ranches were established and cattle were allowed to roam freely, but gradually there was an increase in the demand for wheat as the number of town-dwellers multiplied. Quick ripening varieties which

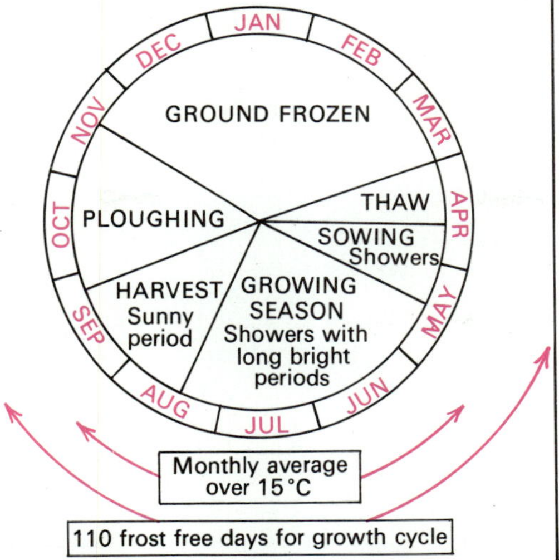

Figure 18.3 Wheat-growing in the Prairies

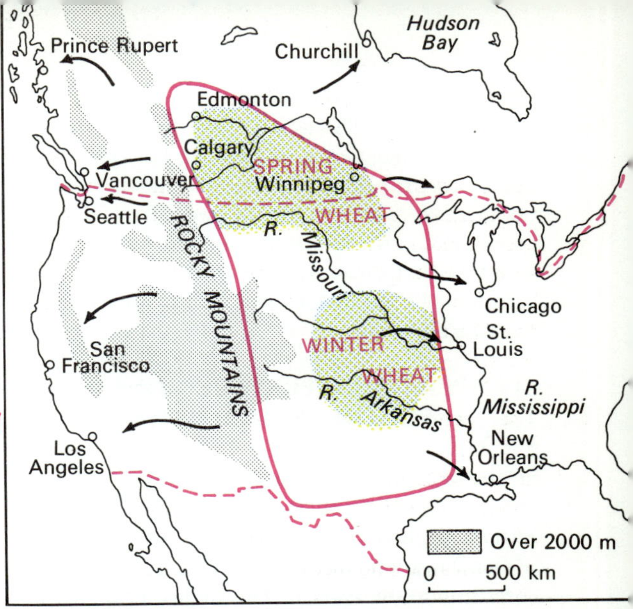

Figure 18.4 The temperate interior region of North America

could mature in 110 days or even less were introduced, and after about 1850 Canadian wheat began to be exported in large quantities to Great Britain, most of that grown in the United States being destined for the growing industrial population in the north-east of that country.

Extensive methods of wheat farming were practised in the prairies long before the steppes. A pattern of huge, monotonous rectangular fields was developed with scattered farm houses many miles apart. Small market towns with a railhead and grain elevators sprang up here and there. One-crop farming (monoculture) led to soil erosion in the drier west, particularly in the United States, but today a more carefully planned agriculture is found. Even in the main wheat-growing areas barley, oats and vegetables have been introduced as part of the crop rotation, and livestock, particularly pigs and cattle, are kept. Soil erosion has been combated by contour ploughing and strip farming. The latter involves the leaving of strips of permanent grass between ploughed land. Irrigation schemes have also been developed near the foothills of the Rockies. In some areas *dry farming* is practised; under this system wheat is grown every other year, the land remaining fallow during the intervening year so that as much water as possible can be stored in the soil.

In the temperate interior of Canada and the northern part of the United States the wheat is spring sown. Farther south maize becomes the dominant crop. This is part of the same 'maize belt' which was mentioned in Chapter 17. The crop is used primarily for the fattening of stock, including both cattle and pigs on the same farm. Farther south still in the temperate interiors we come to a region where wheat is grown as a winter crop, being sown in late autumn, so that it has a longer period in which to mature.

TEMPERATE INTERIOR REGIONS IN THE SOUTHERN HEMISPHERE

Here the summers are hotter and the winters much milder than in the equivalent regions of the northern hemisphere. In the Murray–Darling Basin of Australia sheep raising for both wool and meat has been important since the land was first opened up in the late nineteenth century. *Merino sheep*, which give fine wool and which were originally developed in Spain, do well in Australia, but being small and lean do not bring high prices when sold for mutton. For this reason dual-purpose breeds have been developed, from a mixture of Merino and English mutton stock. Wheat is also important in the southern part of this region, and many of the farms produce both wheat and wool. The grain is hard, and although the yield is low it is of high quality.

The *high veld* of the South African plateau is a region of large farms. In the wetter parts the chief crop is maize which is used for fattening cattle as well as for human food. Most of the drier parts are sheep- and goat-rearing country. There has been considerable mining and industrial development in the vicinity of Johannesburg which has created a large market for farm produce.

Describe the significance of these two scenes, one near Regina and the other on the St Lawrence seaway (*Photos Canada House*)

Describe this scene in the Canadian Prairies. At what time of the year was it taken? Give reasons for your answer. Make a carefully labelled plan of a prairie farm based on this photograph (*Photo Canada House*)

Exercises

1 In middle latitudes continental interiors have extreme climates compared with coastal areas.
  (*a*) Show how this statement is true by reference to the graphs on pages 151 and 157.
  (*b*) Give reasons for the differences.

2 Study figure 18.4.
  (*a*) What is the difference between spring wheat and winter wheat?
  (*b*) Describe and explain the difference in their distribution.

3 (*a*) What is a collective farm?
  (*b*) What is a state farm?
  (*c*) How do these farms differ?

4 (*a*) What methods of protecting the soil against erosion have been used in the prairies and the steppes?
  (*b*) Why are soil conservation methods so important?

5 Study the photograph on this page. Scenes like this have been called a *chequerboard landscape*.
  (*a*) What does this mean?
  (*b*) How did this landscape come into being?

Chapter 19

Cool Temperate Western Margins

Climate

This region lies on the western sides of land masses and extends approximately from latitudes 40° north and south to 65° north and south. Here the prevailing westerly winds blow from the sea to the land throughout the year. Depressions frequently pass over from the sea so that the weather shows a great variability from day to day, though there are no great extremes in climate. The winters are mild and summers cool, with mean temperatures of around 5°C and 15°C respectively, and the change from winter to summer is so slow that it is normal to consider the year as being divided into four seasons—spring, summer, autumn and winter.

Most rainfall is of a cyclonic type and is fairly evenly distributed throughout the year, but the total amount varies considerably from one place to another. For example, in England the total ranges from over 2000 millimetres in some of the hilly western regions where there is heavy relief rain to as little as 500 millimetres on the eastern lowlands.

Natural Vegetation

The cool damp conditions favour the growth of broad-leaved deciduous forest, often with a thick undergrowth. The leaves are shed in winter when temperatures are too low for growth. The trees include the oak, sycamore, beech, ash, elm and willow, and yield valuable hardwoods used for furniture and house interiors. In mountainous areas and on sandy soils coniferous trees may replace deciduous trees, and much of the land may be covered with heather, coarse grasses, gorse and bracken.

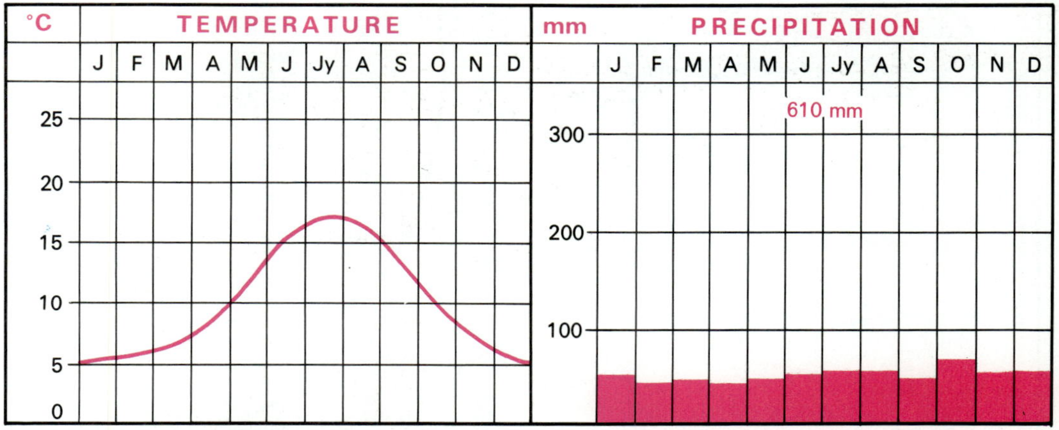

Figure 19.1 Climate graphs for London

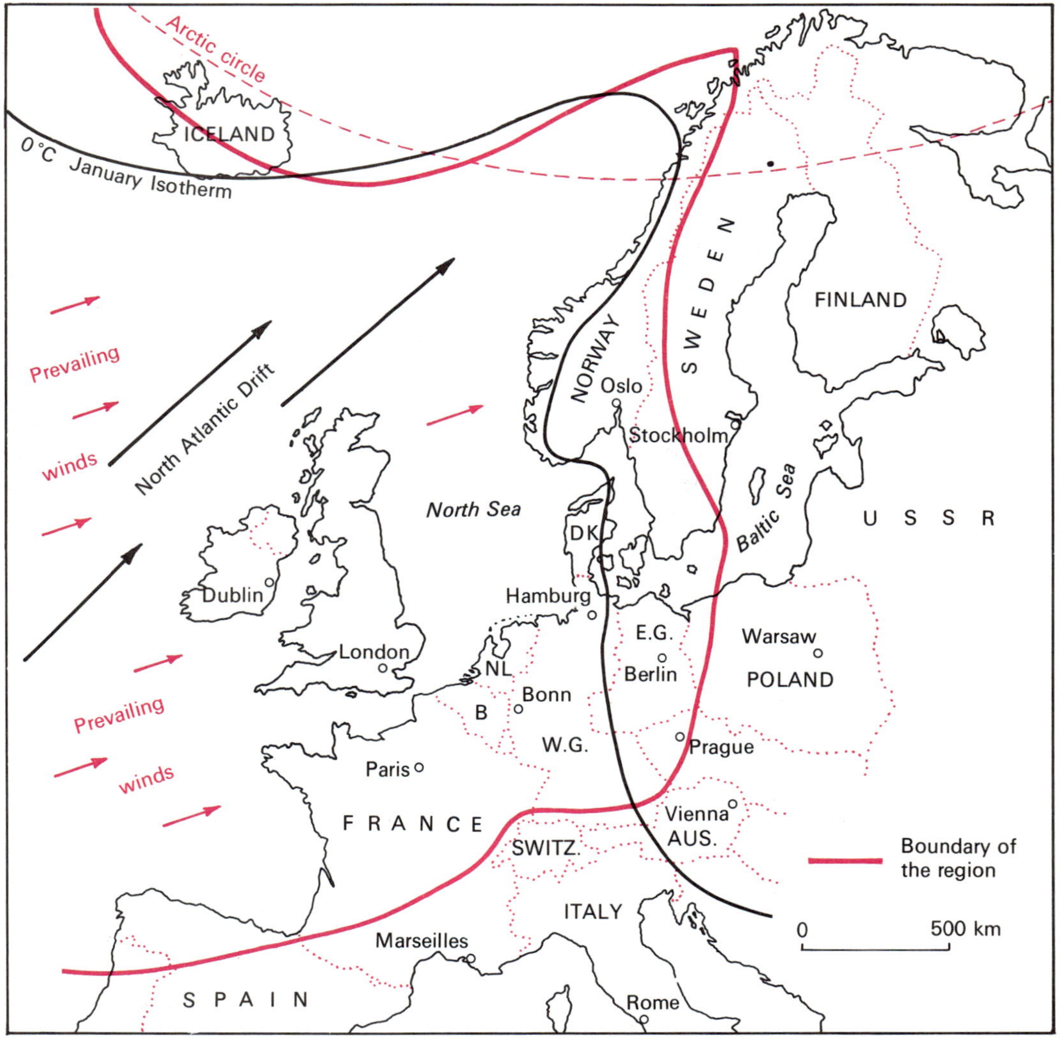

NORTH-WEST EUROPE

The climate of north-west Europe benefits considerably from the influence of the warm North Atlantic Drift. This accounts for the northward extension of the region along the coast of Norway. Another important effect is that harbours well to the north of the Arctic Circle are free from ice.

Some of the most densely populated countries in the world are found here. It was in Britain that the Industrial Revolution began in the second half of the eighteenth century, helped on by many inventions. Today some of the largest manufacturing areas in the world are found in north-west Europe. The history of some of the nations, such as Great Britain, the Netherlands and France, has been a very strong factor. They had set up colonies and built up large fleets of merchant ships. This experience of trading and colonisation was used to develop overseas markets, and these in turn provided raw materials. Another major factor was the existence of large coalfields which provided the power to drive the machines.

A great variety of farming activities is possible in the cool temperate western margins but the dense urban populations have given a distinctive aspect to agriculture. There is a need to provide this market with milk, fresh vegetables and other produce which will deteriorate if not consumed within a short period. These are known as perishable foods. On the other hand many countries including Great Britain must import considerable quantities of non-perishable foods, particularly cereals.

Figure 19.2 Cool temperate western margins of Europe

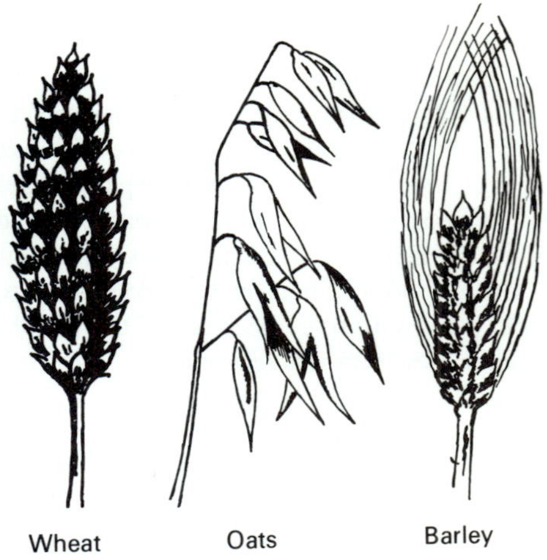

Figure 19.3 Temperate cereals

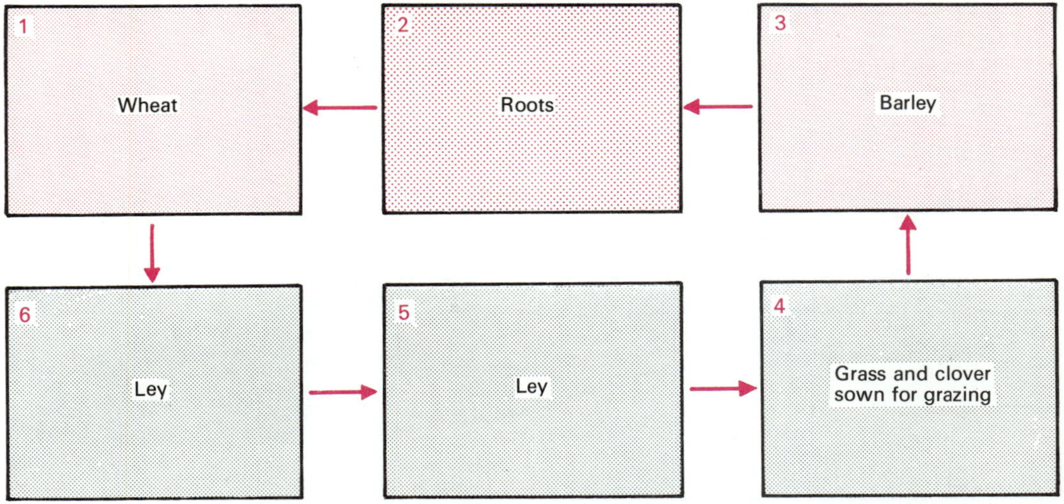

Figure 19.4 A six-year rotation Field 2, now growing a root crop such as sugar beet, turnips or mangolds, next year would grow barley, then be under grass for three years followed by wheat.

Much of the natural vegetation has been replaced by farmland. In lowland areas grassland provides rich pasture for livestock. *Cattle*, which thrive on long, nutritious grasses, are kept here in large numbers to provide fresh milk and meat. In very mild areas they graze out of doors throughout the year, but in other less mild areas they are kept indoors and stall fed for a number of months in winter. Cattle are also kept in some highland areas where there is adequate pasture although here *sheep* are generally more numerous. Sheep do well on the shorter, coarser grass though if possible it should be dry underfoot since they are prone to suffer from foot rot and other diseases when kept on damp land.

To an increasing extent cattle and sheep, as well as pigs and poultry, are kept on mixed farms. Intensive methods are employed wherever possible, much use being made of manure and fertilisers. In order to maintain the soil in good condition, *crop rotation* is normally practised.

There have been many changes in farming, especially in the farmlands of eastern England. Many of the large farms are now owned jointly by companies involved in many other activities besides farming. They operate their farms on a large scale to obtain maximum profits. This kind of farming is called *agribusiness*. It is often accompanied by changes in the type of farming, from mixed farming to arable farming, and by changes in the farming landscape, as hedges are removed to make larger fields. Where such changes occur, crop rotations may be much simplified with almost continuous cropping of cereals, especially barley.

An important aspect of intensive agriculture is *market gardening*. This provides the vegetable and fruit markets with a regular supply of fresh produce. To extend the season hothouses are often used, in particular for salad crops. Orchard fruits include apples, pears, plums and cherries, and soft fruits include raspberries, black currants, gooseberries and strawberries.

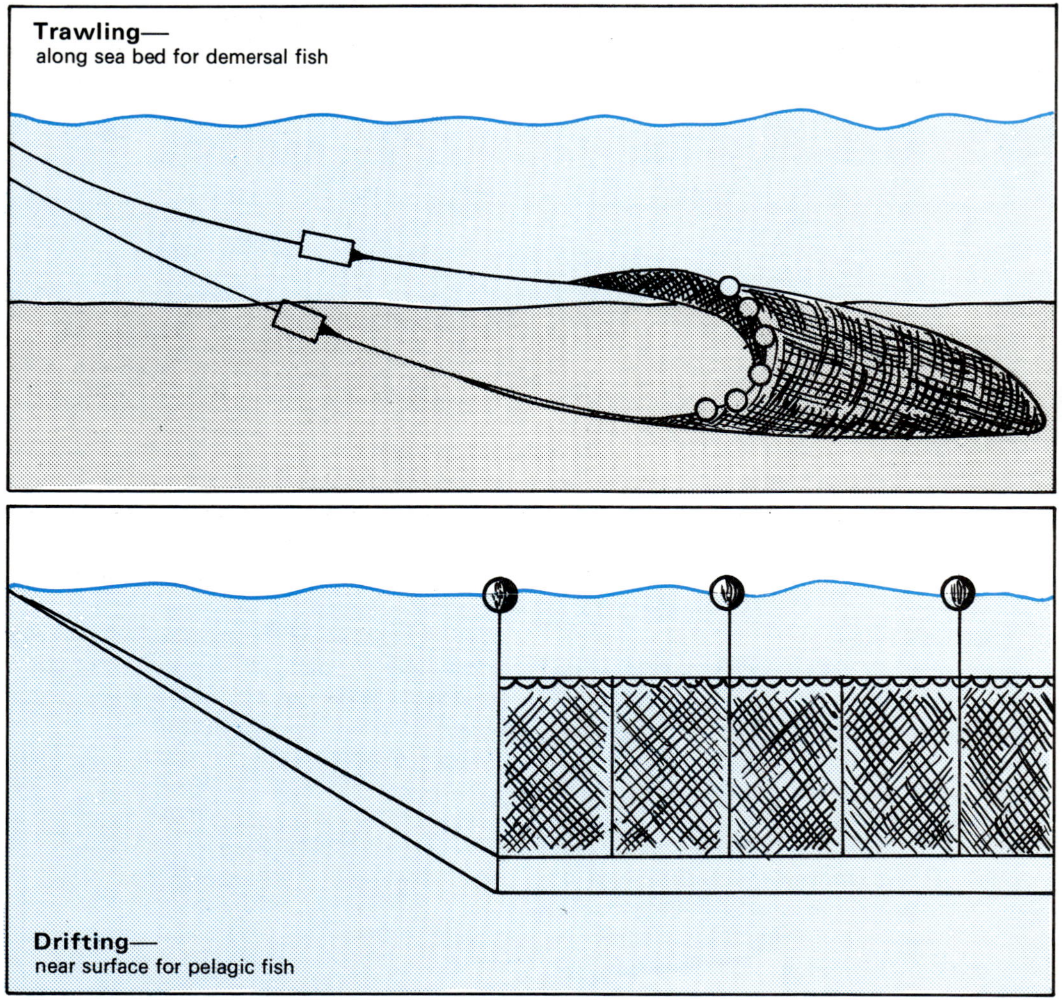

Figure 19.5 Fishing methods

The shallow waters of the continental shelf round western Europe are rich in plankton and provide ideal breeding grounds for fish, whilst the numerous inlets form the harbours needed to shelter small fishing craft. Two groups of fish can be distinguished. *Pelagic fish* which keep to surface or shallow waters, such as, herring, mackerel and pilchards, and *demersal fish*, which are caught near or on the sea floor. These include cod, halibut, haddock, plaice and sole. One-fifth of the world's fish catch comes from this region. Overfishing has resulted in the collapse of some species, such as herring, with others being in danger. Restrictions have been imposed by most countries and by the EEC, either by prohibiting access to fishing grounds within the 320 kilometre limit, or by limiting or banning altogether catches of particular species of fish.

OTHER COOL TEMPERATE WESTERN MARGINS

In the Americas these regions are restricted by great north-south mountain ranges which form a climatic barrier close to the west coast. In *North America*, in British Columbia and the states of Washington and Oregon in the United States, the most extensive lowlands are confined to the valleys of the Fraser and Columbia rivers and their tributaries. Here agricultural activities are similar to those practised in western Europe. Fruit, particularly apples, is a speciality, and in some of the more sheltered valleys even apricots are grown. The Coast Ranges create a rain shadow for some of

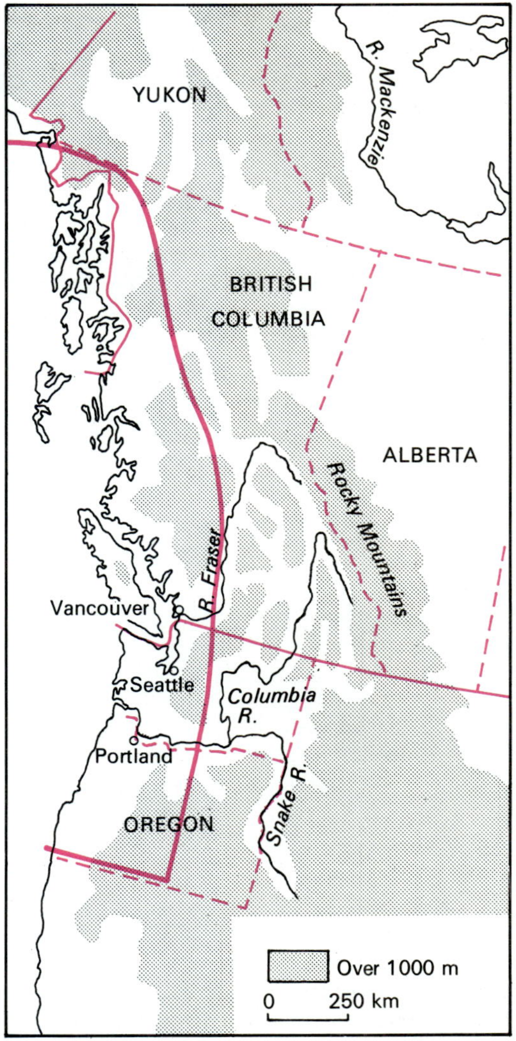

these interior valleys so that the rainfall is often very meagre. Irrigation schemes have been developed here, especially for orchards.

Fishing for salmon and halibut is important, but since the area is remote from large centres of population many of the fish are canned. Some valuable types of coniferous tree, including the giant sequoia and Douglas fir, grow on the mountain slopes, their timber being used for constructional purposes rather than for pulp.

Southern Chile is the least developed of these regions. Settlement has been restricted by the scarcity of flat land and the exceptionally heavy rainfall. The broken coastline with deep valleys, rocky islands and precipitous slopes makes accessibility difficult. There are forests of pine and beech but little lumbering.

In *Australasia* the island of Tasmania has sheep and apple orchards, and South Island, New Zealand, is noted for its sheep products of mutton, lamb and wool. Both these islands are mountainous and most of the settlement and farming is restricted to the broader valleys and coastal plains.

Figure 19.6
Cool temperate western margins of North America

Exercises

1 Explain why the natural vegetation described on page 157 is now found in only a few parts of this region in Europe.

2 In winter the coast of Norway is always free of ice but the east coast of Canada is ice-bound much further south. Why is this?

3 Trace an outline map of western Europe west of a line joining Archangel and Milan, and include part of the North Atlantic Ocean. On the map:
- (*a*) Mark and name the January isotherm for 0°C, the July isotherm for 15°C and the isohyet for 1000 millimetres per annum.
- (*b*) Show with arrows the general directions of the North Atlantic Drift and the prevailing winds.
- (*c*) Compare and contrast the temperature pattern in January and July. Give reasons for the differences you notice.
- (*d*) Describe and explain the distribution of rainfall in the region.

4 Explain the need for irrigation schemes in some of the interior valleys of British Columbia, Washington and Oregon.

Study the four photographs which illustrate contrasting types of farming in the British Isles. Under what conditions does each type take place? Suggest likely areas for each of the scenes (*Photos by courtesy of the Farmer's Weekly*)

174BEW

Chapter 20

Cool Temperate Eastern Margins

Climate

These lie between about latitudes 40° and 50° north and south and have a more extreme climate than the cool temperate western margins. Whilst the winter months have mean temperatures below freezing point the warmest months may average between 15°C and 18°C. The low winter temperatures are a result of the prevailing westerly winds which blow from the cold interiors of the continents. In summer the winds are more variable due to the influence of the continental low-pressure system, towards which the winds blow at times, instead of in the more normal westerly direction.

The precipitation is moderate in amount and is fairly evenly distributed throughout the year, falling as rain in summer and snow in winter. Much of this is brought by depressions, some of which travel right across the continent whilst others, in summer, are drawn in towards the continental low pressure from the bordering seas to the east.

Natural Vegetation

This generally consists of mixed forest including both coniferous and deciduous trees. Coniferous trees, especially spruce and pine, cover much of the higher land, whilst on the lower land oak, beech, ash, maple, chestnut and elm are found.

EASTERN CANADA AND NORTH-EAST UNITED STATES

This region includes the St Lawrence lowlands, Great Lakes area, Maritime provinces and Newfoundland in Canada, and New England, New York and other parts of north-east USA. The coastal areas are greatly influenced by the cold Labrador Current which keeps the St Lawrence and its estuary icebound for several months of each year.

Figure 20.1 Climate graphs for Quebec

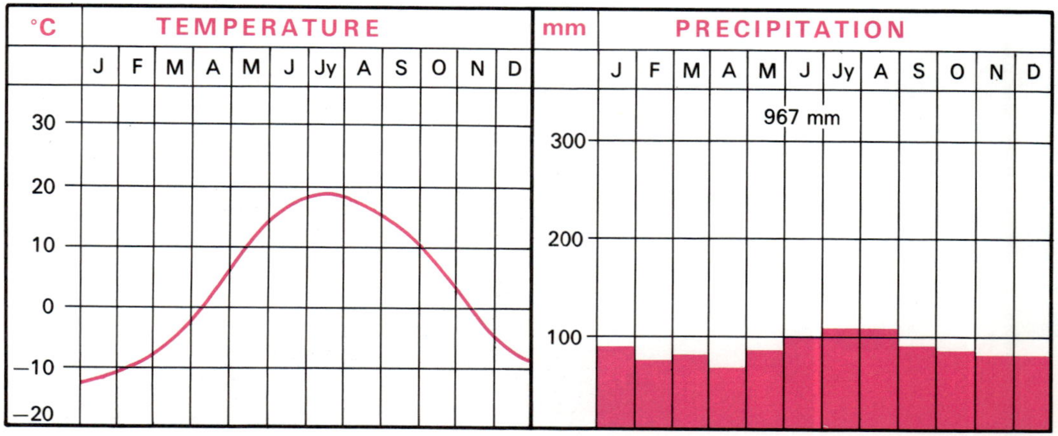

Figure 20.2 The St Lawrence lowlands and north-east USA

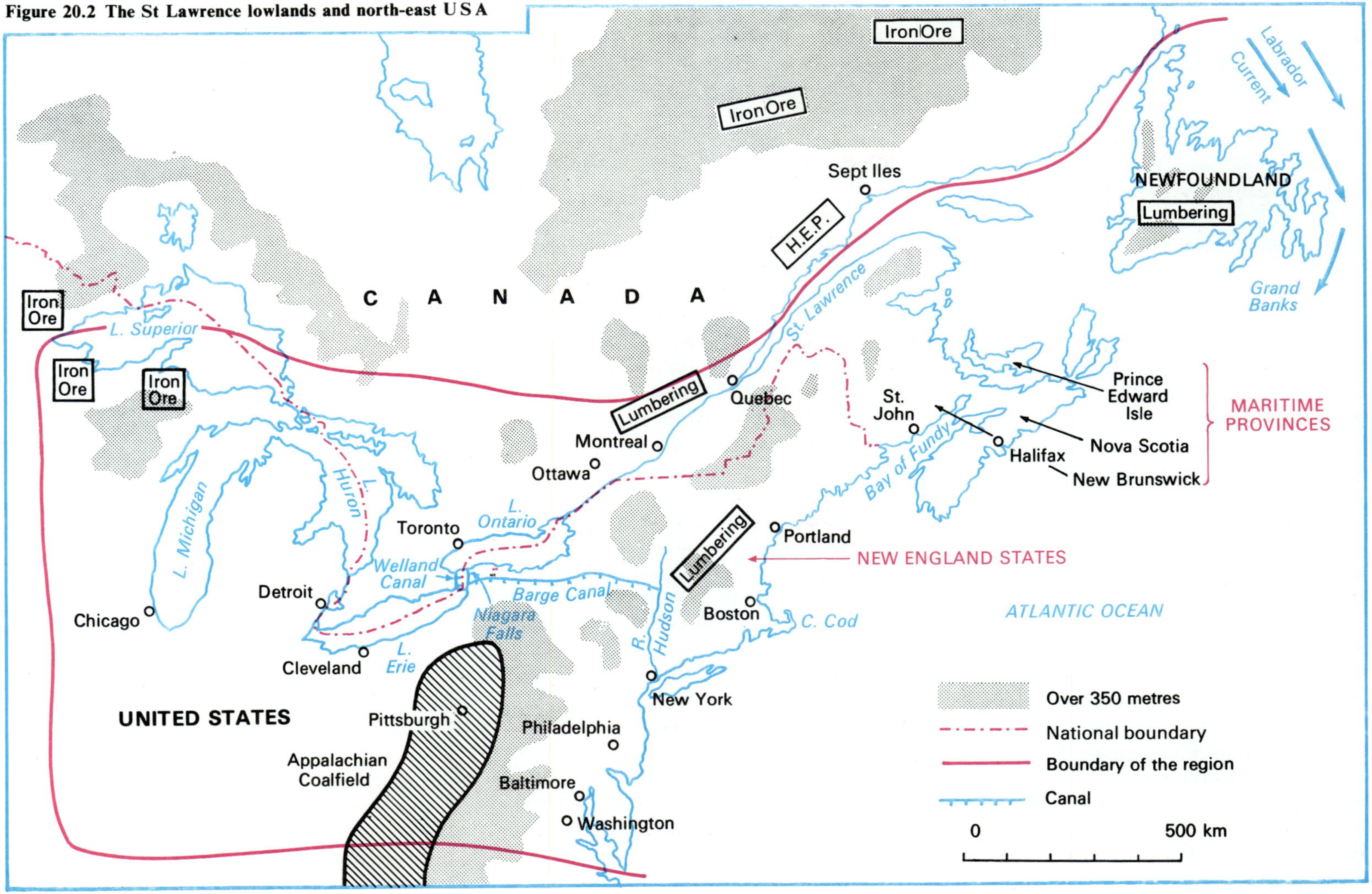

In the forests much of the good timber has been cut including the valuable white pine. Lumbering still continues in the highlands of New England, along the northern shores of the St Lawrence and in Newfoundland where spruce has for long been cut for newsprint.

Much of the lowland is devoted to mixed farming with cattle, hay, wheat and oats as the principal products. The climate is rather wet for wheat which can be produced more cheaply in the prairies.

Dairy cattle supply fresh milk to the local industrial towns, and a further development reflecting this large population is the market gardening which is concentrated on the lighter sandy soils along the coastal tracts of the USA. In North America this is known as *truck farming* since the fresh vegetables and fruit are taken directly to market by motor truck. Some areas specialise in fruit, for example apples in parts of Nova Scotia and cranberries around Cape Cod. Potatoes are produced on the rich soils around the northern end of the Bay of Fundy. The St Lawrence lowlands are inhabited mainly by French-speaking Canadians who pursue traditional methods of mixed farming on smallholdings.

As in north-west Europe the shallow waters of the continental shelf have provided valuable fishing grounds. The area of the *Grand Banks* has long been an important fishing ground. The cod fisheries of the Grand Banks were the best-known in the world, but catches of cod and other important fish dropped markedly as a result of overfishing. With the introduction of Canada's 320 kilometre fishing limit in 1977, excessive exploitation has been limited. A national fishing plan has been developed, one aim of which is to limit catches while stocks recover.

Traditionally, fishing in the area was carried on from small settlements around the rugged coast using small craft called *dories*. Even today the majority of fishing boats are less than 12 metres in length.

The many natural harbours along the coast have provided the sites for Halifax, St John, Portland (Maine) and Boston.

The prosperity of the region is based on its manufacturing industries, the growth of which was aided by the great natural assets of the area. The Appalachian coalfield, the ironfields around Lake Superior and in Labrador and Quebec are linked by the Great Lakes and their associated waterways.

Navigation on the Great Lakes and St Lawrence has been made possible by the construction of canals by-passing the many rapids, including the Soo Canals between Lakes Superior and Huron and the Welland Canal which avoids the Niagara Falls, between Lakes Erie and Ontario. In 1959 a series of deep-water canals known as the St Lawrence Seaway was completed between Lake Ontario and Montreal, so that ocean-going ships can now reach all the main ports on the Great Lakes.

The importance of the region is underlined by the location here of America's greatest city, New York, and the federal capital, Washington, as well as Canada's two leading cities, Montreal and Toronto.

MANCHURIA AND EASTERN SIBERIA

This region shows some interesting differences from the North American region. The winters are longer and more severe and the summers shorter and hotter. There is very little precipitation in winter, reflecting the dominance of the interior high-pressure system which develops at that time of year.

Physically Manchuria is a lowland basin fringed by mountains so that continental influences are even more marked here. Much of the land is devoted to the cultivation of cereals which include wheat, maize and millet. One speciality of the region is the giant millet known as *kiaolung*. This grows to over 3 metres in height, and its grain provides food, the leaves a valuable fodder and the stalks are used for fencing, thatching and firewood. The chief money crop is the *soya bean* which can be used as a vegetable, and is a source of oil for cooking, lighting and the manufacture of soap and paints. It is also valuable in any rotation system with cereals as it is a nitrogenous plant.

Recent years have seen a great change in the economy of Manchuria. During the first half of the twentieth century the population increased five-fold, as many millions of Chinese immigrants arrived in search of employment and land. Rich seams of coal and nearby deposits of iron ore have provided a basis for an expanding iron and steel industry, and hydro-electricity has been developed as an alternative source of power.

Parts of eastern Siberia are also included in this region. This area has until recently been thinly populated, but efforts are now being made to develop its large reserves of timber and minerals. Russia's principal Far Eastern port, Vladivostok, lies here. It is the terminus of the Trans-Siberian Railway, but its harbour is frozen for several months in winter, owing to the influence of the cold Kuriles current. There are valuable fishing grounds off the east Siberian and northern Japanese coasts.

PATAGONIA

This part of Argentina is very dry, deteriorating to semi-desert with scrub vegetation. Elsewhere there is poor grassland. The region is in the rain shadow of the Andes so that the prevailing westerly winds bring very little rainfall. The main occupation is sheep farming, and wool and small quantities of mutton are produced.

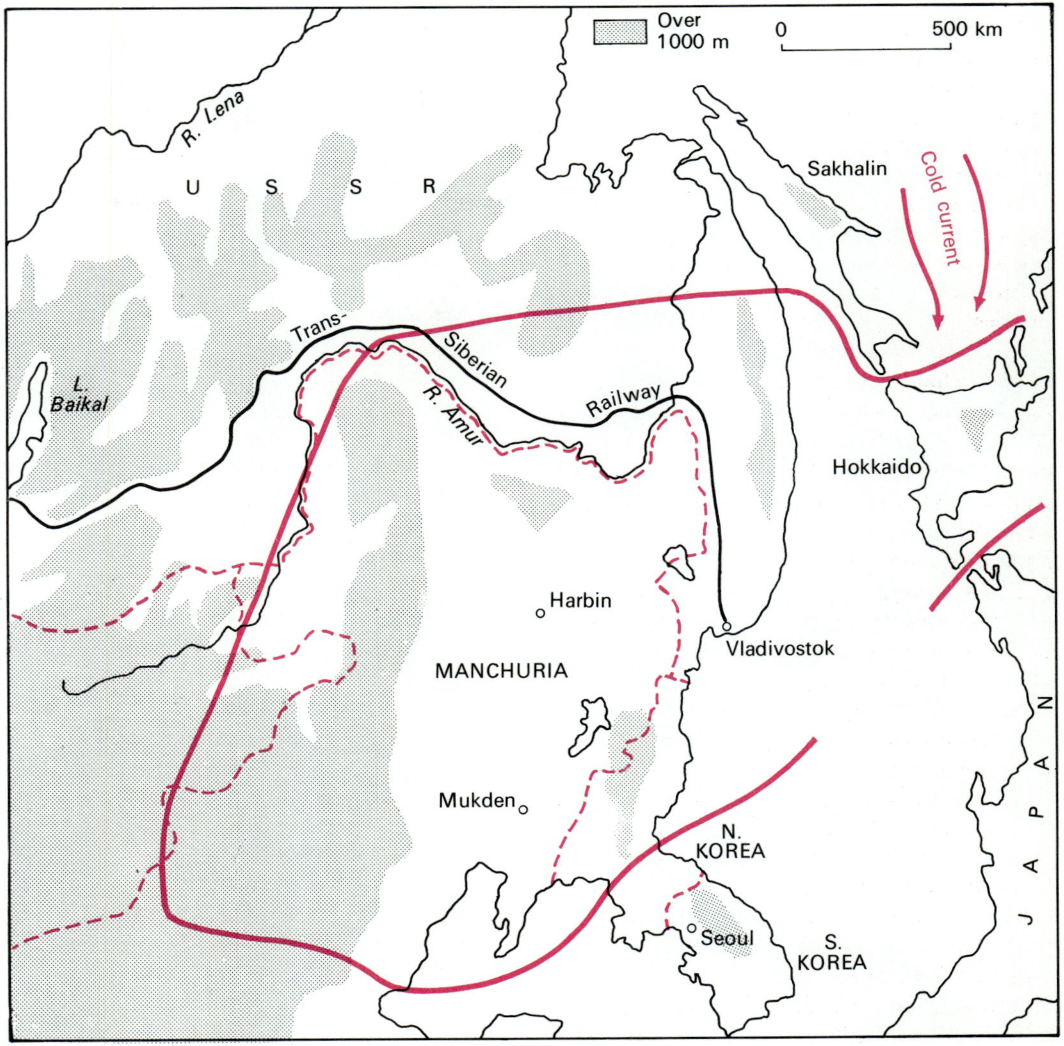

Figure 20.3 Manchuria and eastern Siberia

In what ways is the scene on page 168 typical of the cool temperate eastern margins? (*U S Information Service*)

This scene in Minnesota illustrates an important soil conservation method. It shows alternate strips of maize and hay with grazing for dairy cattle. How does it conserve the soil? (*U S Information Service*)

Exercises

1 Study the climate graphs given in this book for London and Quebec and answer the following questions:
- (*a*) What are the mean temperatures for the coldest months and for the hottest months for the two places?
- (*b*) Compare the annual rainfalls in respect of the total amounts and the distributions throughout the year.
- (*c*) Give as many reasons as you can for any differences you have noted between the climates of the two places.

2 Study the paragraph on page 166 referring to *truck farming*.
- (*a*) What is meant by the term *truck farming*?
- (*b*) What environmental advantages do these truck farming areas have?
- (*c*) In what way is the location along the east coast an advantage?

'The environment in Patagonia is so limiting that sheep farming is the only possible type of farming'. Explain this statement.

Chapter 21

Cold Temperate Regions

Climate

The distinctive features of the climate of these regions are the long severe winters and short mild summers. For six months of the year the temperature rarely rises above freezing point. During this time precipitation is in the form of snow. The total snowfall is not great, but the snow remains on the ground until the following spring. Light showers of rain fall during the summer months. The total precipitation throughout the year is normally equivalent to between 250 and 375 millimetres of rainfall.

Natural Vegetation

Although the climate is dry the loss of moisture by evaporation is small and there is sufficient for tree growth. However, in order to survive, trees must have the ability to withstand the rigours of winter and at the same time make sufficient growth during the relatively short growing period. *Conifers* are suited to these conditions since they have thick bark and resinous, needle-like leaves which serve to restrict transpiration. They are evergreens so that they do not have to produce new leaves each spring and consequently can make a quick start in growth with the spring thaw, which provides moisture from the melting snow. The seeds form within the cones, which serve to protect them from rodents. The flexible branches bend beneath the weight of snow, causing it to slide off. In a coniferous forest the trees grow close together, undergrowth is almost absent and the ground is covered with a carpet of fallen needles.

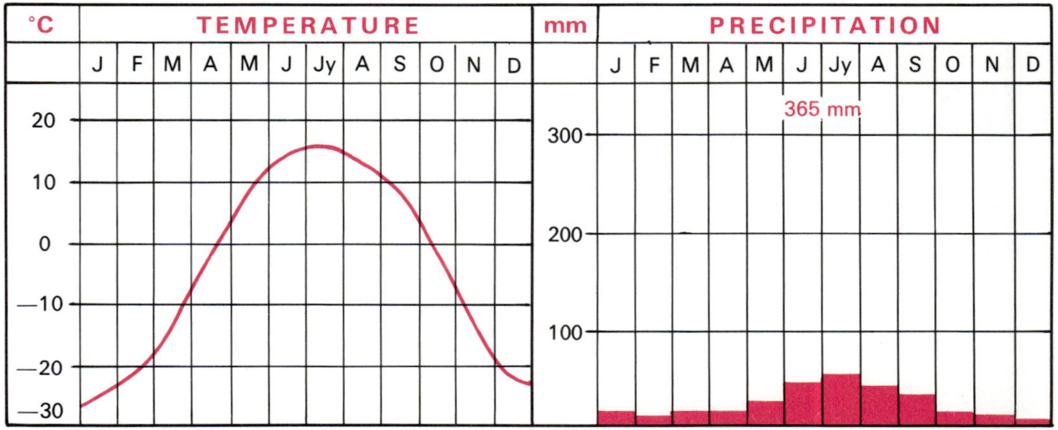

Figure 21.1 Climate graphs for Fort Vermilion

Most of the trees are *softwoods* and include many varieties of pine, spruce, fir and larch, the latter being a deciduous tree. Where conditions are not quite so severe some deciduous hardwoods are found, for example poplar and maple, whilst another, the birch, can survive even in the north of the region.

Figure 21.2 The coniferous forest belt of Canada

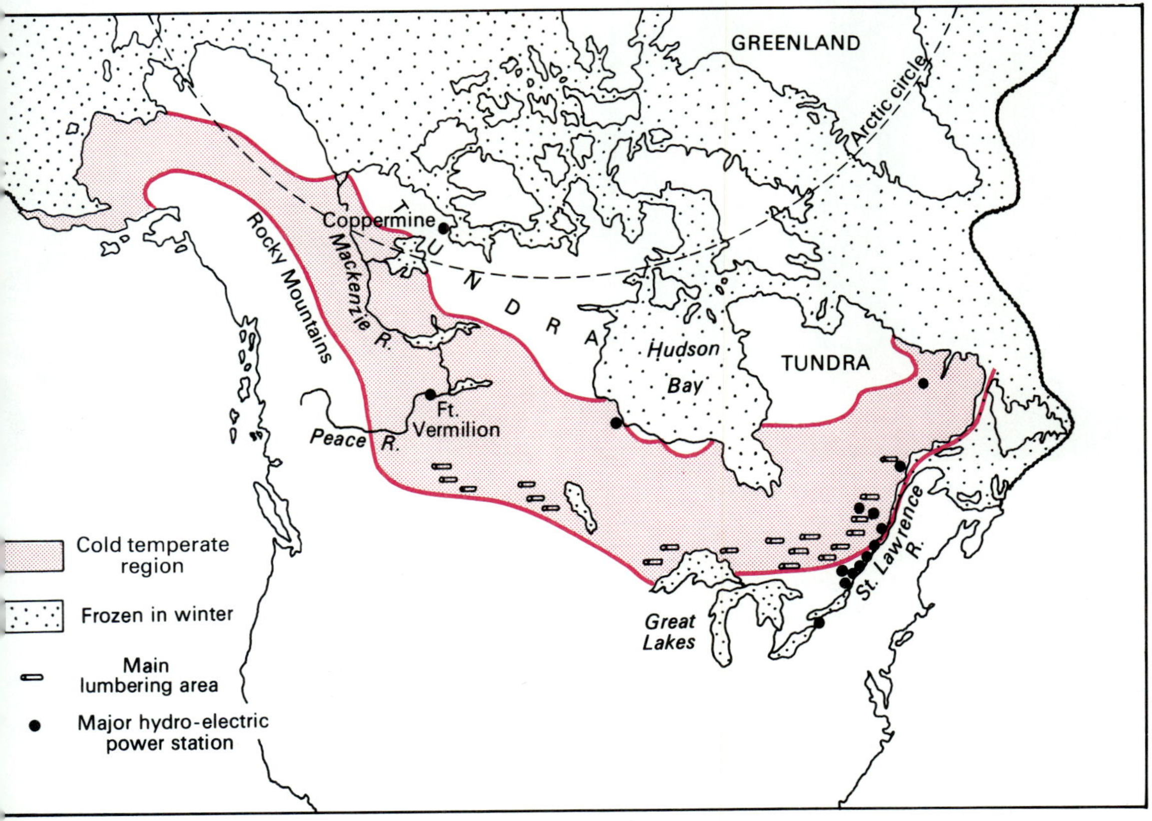

THE CONIFEROUS FOREST BELT OF CANADA

Pioneers were first attracted to these great forests for the *fur-bearing animals* rather than for the timber. Trading companies, which included the famous Hudson's Bay Company, established trading posts to purchase the pelts from native Indian, Eskimo and European trappers. The pelts of wolves, foxes, ermine, mink, marten, beaver, musk rat and bear were in great demand, but today the number of fur-bearing animals has dwindled and they have retreated to the more inaccessible northern parts of the forest. There is still some trapping, but large quantities of furs are now produced by breeding the animals in captivity on fur farms. The silver fox, beaver, mink and musk rat in particular are bred to provide coats and other garments for the fashion-conscious women of cool temperate lands.

Lumbering has been confined to the more accessible parts of the southern forest belt. The first areas to be exploited on a large scale were just north of the St Lawrence River. Here a definite seasonal pattern of industry has been established. The timber is cut during the winter when the sap is not rising so that tree felling is easier. The snow facilitates haulage to the banks of the frozen rivers, which provide a swift and simple means of transport for the logs in spring. The saw-mills and pulp-mills are located downstream from the cutting areas, often near a point suitable for generating hydro-electric power. Thus the streams are of value not only as a form of transport but as a source of power.

Softwoods have a wide variety of uses, ranging from railway sleepers and telegraph poles to matches and plywood. Some timbers, such as the Douglas fir and white pine, are valued particularly in the building trade for floorboards, doors and window frames.

Large quantities of wood are consumed in the manufacture of *paper*. Spruce provides the chief source for this. The pulp and paper mills are located primarily on the north side of the St Lawrence River and Great Lakes. Here the logs are cut into chips, ground into shreds and then mixed with chemicals so that the fibres are broken down into a pulp. This is then washed, bleached, rolled and pressed. The biggest demand on Canadian paper mills is for newsprint, and large quantities of this are exported to the United States and Great Britain.

In addition to timber and paper a large number of very useful substances including cellulose, resin and turpentine are obtained from softwoods. Apart from these a valuable synthetic fibre, rayon, is produced from wood pulp.

The cold temperate climate is generally unsuitable for agriculture and the soils on cleared forest land are often lacking in the nutrient elements needed for crop production. However, in certain southern areas such as the Peace River district there are more fertile soils where quick-maturing varieties of grain, including wheat, are grown, as well as hay and root crops. The last two serve as fodder for cattle. With the growing exploitation of Canada's minerals many mining settlements have grown up which provide markets for local farm produce.

Tree felling (*Bowater Paper Corporation Ltd*)

Logging (*U S Information Service*)

Study the five photographs and then write a full account of the various processes involved in the production of paper, beginning with the actual felling of the timber.

Log boom at Corner Brook (*Bowater*)

Logs stripped of bark ready for grinding into pulp (*Bowater*)

Paper mill (*Bowater*)

THE CONIFEROUS FOREST BELT OF EURASIA

The cold temperate region here covers a much greater area than in Canada and in the interior the climate is even more extreme, especially in Siberia. The region having the coldest winters in the world is in north-east Siberia, in the neighbourhood of Verkhoyansk. It is often referred to as the 'cold pole' and has a mean January temperature of about −50°C.

Figure 21.3 The coniferous forest belt of Eurasia

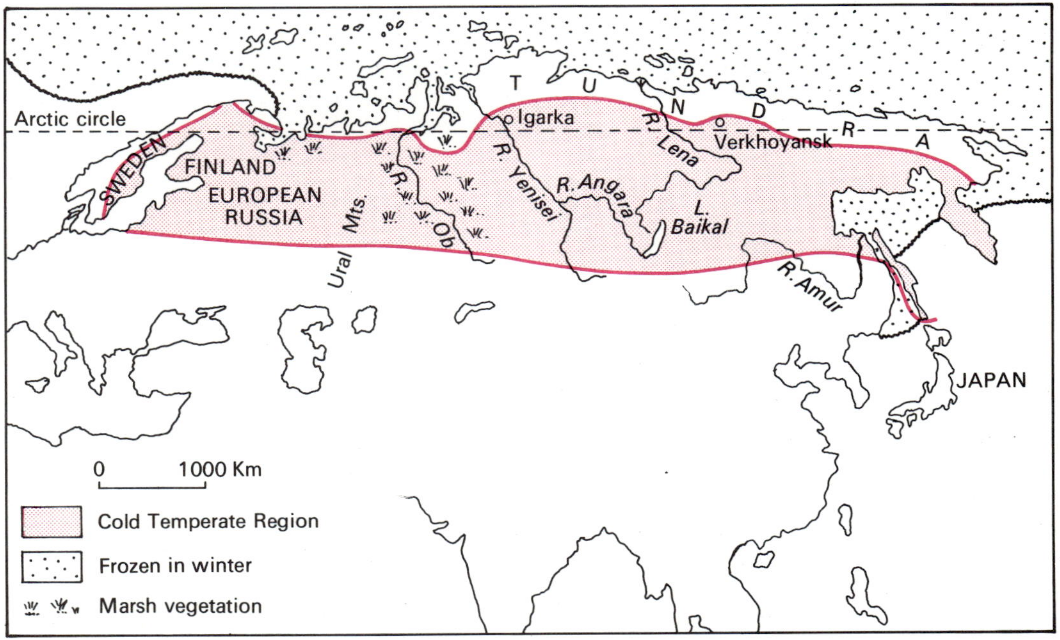

The Siberian coniferous forest belt is known as the *taiga*, and contains enormous reserves of timber, but lumbering activities here are greatly hampered by physical conditions. The vast rivers drain northwards, and since the spring thaw begins in the south the meltwaters flow north on to the still frozen land, causing extensive flooding. Consequently the timber is frequently of poor quality. Transportation is difficult because the Arctic ports of the USSR are frozen up for much of the year, though the 'Northern Sea Route' is kept open by icebreakers for a few months in summer.

Farther west in Sweden, Finland and European Russia, the timber is more easily obtainable, and a larger proportion of the forests has been cleared. The governments of these countries now carry out schemes of planned afforestation.

In Siberia many new towns have been created, most of them adjacent to rich mineral deposits, coalfields or on rivers where hydro-electric schemes have been developed. One of the first of these developments was on the Angara River, north of Lake Baikal, which attracted engineering, chemical and timber industries. With more developments other industries which use large amounts of power, such as aluminium smelting and timber-based chemicals, have been attracted. To increase the local food supply, experiments are being made to widen the range of crops which can be grown on scattered clearings in the taiga. The small rural population hinders agricultural development. As in Canada, trapping has declined and fur farming is on the increase.

Exercises

1 Over-exploitation of some forest resources in Canada and the USA pointed to the need for conservation measures.
   (*a*) What is meant by over-exploitation?
   (*b*) Suggest what measures were adopted.

2 Over one-third of all the USSR's timber production comes from Siberia and the Far East.
   (*a*) What environmental problems have to be overcome?
   (*b*) In what way is distance from the main population centres a problem?

3 Draw a flow chart to show the various stages or steps in the production of either furniture or paper. Label your flow chart fully.

Chapter 22

High Mountain and Tundra Areas

HIGH MOUNTAIN AREAS

A glance at the world physical map in your atlas will show that a large proportion of the land masses lies at an elevation of 1000 metres or more above sea level. The climatic conditions in these areas vary considerably with both the altitude and the latitude. Mountains have lower average temperatures than the surrounding plains, since temperatures decrease with height, but whereas in tropical areas a height of 2000 metres produces a temperate climate attractive to man, in temperate areas the same height produces a frigid climate, unsuitable for human settlement. The aspect of a mountain slope can also have a marked effect on temperature figures; for example in the northern hemisphere, particularly away from the Tropics, south-facing slopes will experience the Sun's rays for many hours during the day whilst north-facing slopes may receive hardly any direct sunlight. The effect of altitude on the precipitation is also most marked. At first the precipitation increases with height, but in the case of very high mountains it reaches a maximum and then decreases. The rain shadow effect whereby windward slopes receive more rain than leeward slopes was explained in Chapter 4.

Distinct vegetation belts can be recognised on mountain slopes. In temperate latitudes broad-leaved deciduous forest is generally found on the lower slopes, giving way to coniferous forest higher up. Above the forest are the Alpine meadows which extend up to the snow line. These meadows are snow covered in winter. In tropical latitudes the lower slopes are often clothed in dense evergreen forest, which is succeeded at greater heights by a savanna type of vegetation. This changes to grassland and scrub as the snow line, which is at a much greater height in these latitudes, is approached.

Figure 22.1 Vegetation belts in mountain areas

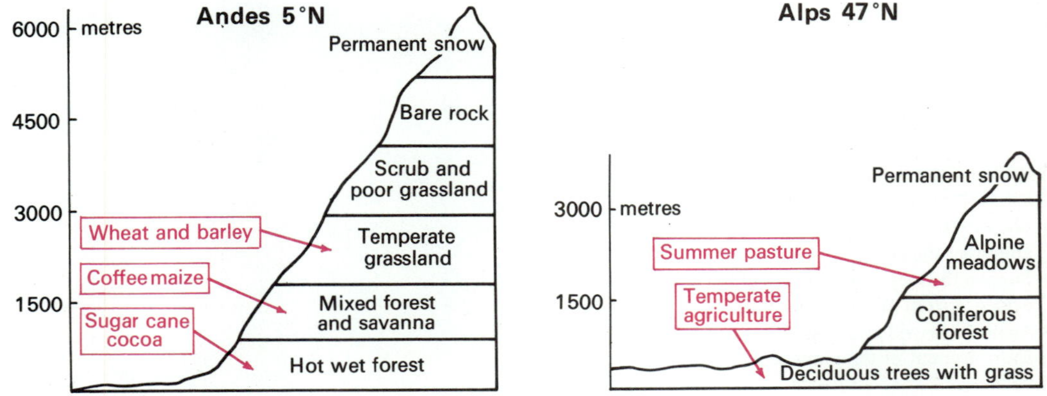

A valley in the Austrian Alps (*D Jones*)
There are several zones: valley floor, wooded slopes, alpine pastures and mountain peaks

a) Identify the zones. Draw a sketch section of the valley and label the zones.
b) Suggest the uses to which each of the zones is put. Use evidence from all the photographs.
c) Again using evidence from all the photographs, write a short account of the main economic activities in this valley.

1 **High alpine pasture**

2 **Haymaking**

3 **Hotel**

4 **Mountain refreshment hut**

(*Photos D Jones*)

In India, China and Japan there has been a movement of peoples from the overcrowded lowlands to find living space on higher land where the forests had to be cleared and slopes terraced. Here we had peoples moving into a more difficult environment. In contrast some mountainous areas may be more hospitable than the adjoining lowlands. The high plateaus of the Andes have invited settlement in the past whereas the Amazon low lands to the east have repelled it. Today these hardy mountain peoples keep llamas, alpacas and vicunas and grow potatoes and other crops. Here in the Andes is another greater incentive to development—the attraction of minerals, including the great tin deposits of Bolivia and the copper and lead of Peru.

More recently the scenic attractions and possibilities for climbing and winter sports have greatly increased the importance of many mountain areas. The tourist industry is now one of the world's greatest employers of labour and is destined to become increasingly important as standards of living rise.

Mountain regions have been settled for a variety of reasons. In the past some of them have provided a refuge for peoples forced to flee from the more fertile lowlands, as in the case of many of the Celtic peoples of Wales. The Alps were also peopled in this way, and here the valley floors provided suitable settlement sites. The inhabitants adopted the system of transhumance, driving their cattle and sheep to the Alpine pastures in summer and bringing them back to their valley farms in winter.

TUNDRA AREAS

Climate

The tundra regions of the northern hemisphere lie within or near the Arctic Circle. Here the winters are long and bitterly cold. During November, December and January there are very few hours of daylight, and about the time of the winter solstice (December 21) there is continual darkness within the Arctic Circle. The summers are short and cool and there is almost continual daylight, although the Sun is never very strong as it is always low in the sky.

Natural Vegetation

For nine months of the year the ground is frozen and snow covered, and even in the summer remains frozen just beneath the surface, a condition known as permafrost. The climate is too cold for the growth, and the natural vegetation consists of mosses and lichens which remain alive beneath the snow in winter, and flowers and grasses which spring up from bulbs and seeds which have lain dormant during the winter. In addition a number of small shrubs are found, including the cranberry and bilberry.

Figure 22.2 Climate graphs for Coppermine

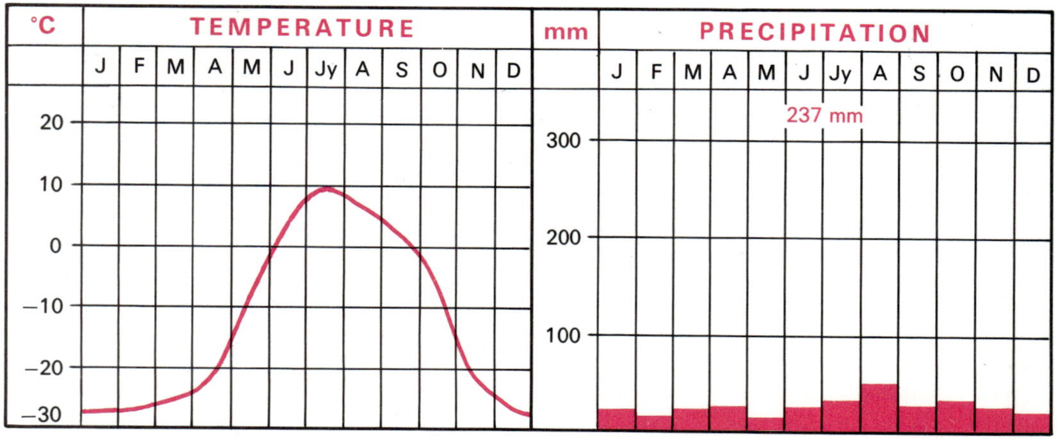

THE TUNDRA OF NORTH AMERICA

Most of the inhabitants of the tundra region in Greenland, Alaska and Canada are *Eskimos*. They formerly lived mainly by hunting and fishing and developed a highly organised nomadic life in difficult surroundings. Their chief prey was the seal which provided meat, blubber, skin for clothing, bones for implements and sinews for thread. For transport they used the sledge, drawn by dogs, on land and the kayak on water.

Virtually no Eskimos follow this traditional nomadic way of life now. Most live in small villages in prefabricated timber houses. The settlements have a school, a medical centre and administrative offices. There is considerable variation in economic activity. In some areas people work in mining or hold other permanent jobs in commercial fishing or in fish-freezing plants. In other areas the herding of reindeer and musk-oxen has been developed, and craft industries and tourism are also important. Even hunting has changed, with motorboats being used on water and snowmobiles on land.

The region itself is important to outsiders for several reasons. Many minerals, including oil, are found there, and chains of radar stations provide North America with warning of missile and aircraft attack across the polar region. Tourism, although more developed further south, is gradually expanding, especially with the development of hunting lodges.

Oil exploration in northern Alaska
Suggest some of the difficulties of developing resources in this environment. (*British Petroleum Co. Ltd*)

THE EURASIAN TUNDRA

In a wide belt extending from northern Norway to the Kamchatka peninsula, reindeer-rearing is still carried on. It is closely linked to the hunting of fur animals, fishing and the catching of seals. The reindeer is a hardy animal and lives off the tundra vegetation. In winter it scrapes away the snow with its forehooves in order to get at the underlying moss. Herders such as the Lapps of northern Scandinavia and the Samoyeds in Siberia follow their wandering reindeer herds from the fringes of the forest in winter to well within the Arctic Circle in summer.

Mining has led to an increase in population in particular areas. Valuable mineral resources have been exploited within the tundra and in the coniferous forest region immediately to the south. Coal, oil and natural gas has been developed in the Pechara Basin, and this has needed the growth of new towns. The West Siberian oil and gas field, which is now linked by pipeline to Western Europe, is an area where very difficult problems presented by the physical environment had to be overcome. The opening up of the Northern Sea Route, and the growth of new ports such as Igarka on the Yenisei, has also encouraged development. To provide the population with fresh food, glasshouse cultivation has been established near mining and transport settlements.

SOUTHERN HEMISPHERE AREAS

In the southern hemisphere the tundra is restricted to a number of islands in the Southern Ocean and to portions of peninsulas projecting from the land mass of Antarctica. Whaling stations have been established on some of these, for it is in these southern waters that most whales are caught today. There are also meteorological and scientific stations both here and on the Antarctic continent itself, which is a vast ice desert experiencing intense cold and strong winds.

GENERAL EXERCISE ON THE NATURAL REGIONS

Study the following table and then (*a*) describe the main features of the temperature and rainfall of the four stations, (*b*) name the natural region within which each station lies; give reasons for your choice.

J	F	M	A	M	J	JY	A	S	O	N	D
					Station 1						
12	13	16	20	24	27	27	27	25	23	18	13
5	5	5	3	3	0	0	0	0	3	3	5
					Station 2						
19	22	27	30	31	29	29	28	28	27	23	19
8	28	36	48	145	302	318	323	251	107	18	5
					Station 3						
−12	−10	−4	5	15	20	22	21	14	6	2	−8
23	20	18	28	28	38	43	33	31	38	36	41
					Station 4						
10	11	13	15	17	22	24	25	23	19	15	12
86	71	69	48	28	15	5	10	46	81	84	94

The upper line for each station shows the temperature in degrees Celsius and the lower line the rainfall in millimetres.

High density development in Hong Kong (*P A Wood*)

PART FIVE

Population and Industry

Chapter 23

Population and Development

POPULATION

The world's population is distributed very unevenly. *Population density*, or the number of people per square kilometre, varies enormously. There are areas which are virtually uninhabited and other areas which are extremely crowded. In fact, 70% of the world's population live on less than 10% of the land surface.

There are several major concentrations of population. These are found in South and East Asia, Europe and North-East USA. Smaller concentrations are found in the lower Nile valley in Egypt, West Africa and coastal regions of South America. The reasons for the existence of these concentrations of people are complex. They are all regions which are able to support large numbers of people either by agriculture or industry. The industrially-based regions generally had the resources needed for industrial growth, especially iron and coal. The agriculturally-based areas have been able to support large numbers partly as a result of favourable climate, soils and relief, but also because highly productive systems of farming were developed. All these densely populated areas have major cities.

Just as the more favourable areas have attracted people, so the less favourable areas have been avoided. The areas that are almost empty of people can be regarded as *hostile* environments which create particular problems for settlement. In high latitudes are the areas with cold climates which have very short summers and long, bitterly cold winters. Apart from small numbers of people living by traditional means, such as reindeer herding or hunting and fishing, the only real concentrations of people are found in mining towns. Where valuable minerals are found, the hostile environment is faced because of the financial rewards. Modern mining settlements, however, have far more comforts than the frontier mining camps of the last century. Arid and semi-arid areas are lacking in the vital necessity of life, water. Concentrations occur only where water is available, whether it is to support agriculture or a city. Mountain regions are generally empty because of the steep slopes, thin soils, lower temperatures, and, at extreme altitudes, lack of air. Some of the largest areas of tropical rain forest, like the Amazon rain forest, are extremely sparsely populated because of poor communications, disease, flooding risk and generally infertile soils.

Figure 23.1 Growth of world population

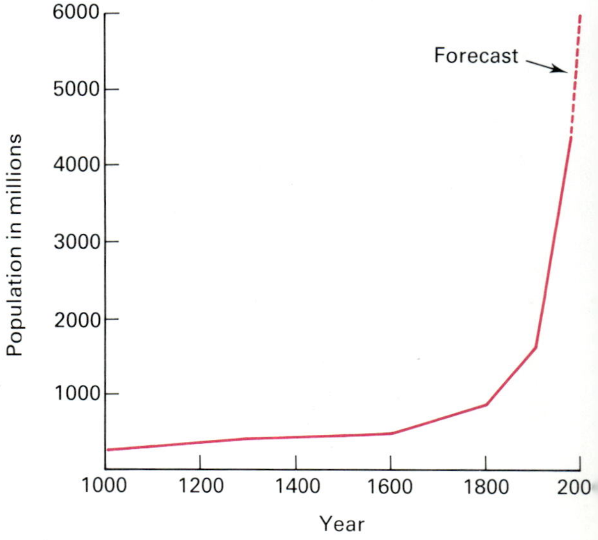

Figure 23.2 World population distribution

Note that this map has been drawn using a different map projection from earlier world maps in the text. This map projection does not distort the areas of continents as much. It is, therefore, better for comparing sizes of empty or crowded parts of the world.

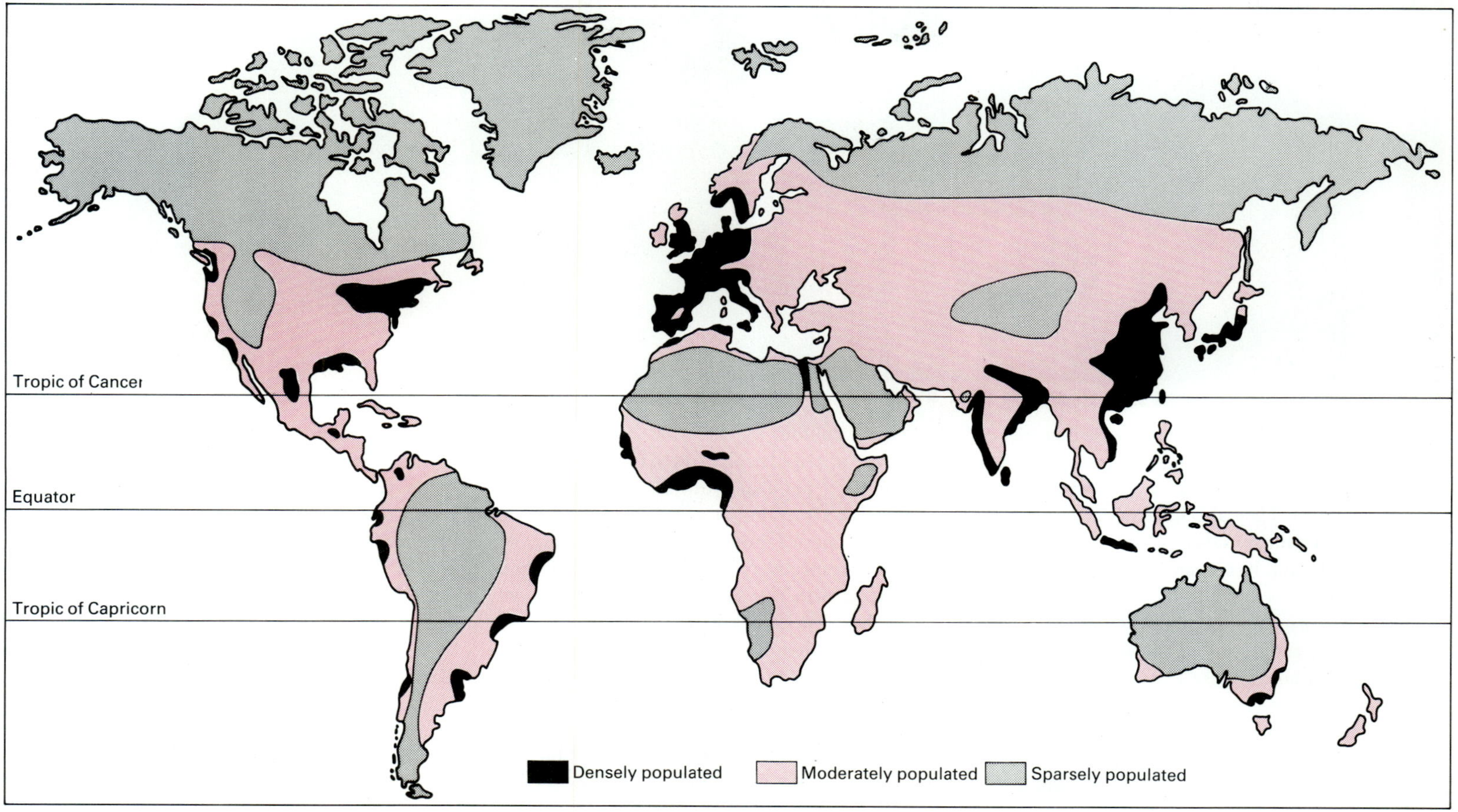

Extremely high density residential area, Hong Kong (*P A Wood*)

Almost completely uninhabited area, Australian desert (*L J Kimpton*)

The areas of the world with moderate densities of population are generally agricultural, but their populations vary considerably according to the farming systems. They also have pockets of denser concentration where industry has developed or where population has been attracted for some other reason.

The rate of *growth* of population is an important element in the geography of the world as a whole and of individual regions. The population of the world has increased rapidly during this century. It took until the year 1500 to reach 450 million, then increased to reach 1650 million by 1900, and by 1980 4432 million. By the end of the century it will reach 6000 million.

The *rate* of population growth for the world as a whole is 1.7% per year, but that figure hides wide differences between major world regions. In Africa and Latin America, population is growing at 2.9% and 2.5% respectively. In North America, however, the growth rate is 1% and in Europe only 0.4%. These variations in growth mean that the actual distribution of population is changing, with some regions being likely to double their population in 25 years.

Population growth or decline depends on the birth and death rates. If the number of births is greater than the number of deaths, the population will grow. Birth rates and death rates are usually written as the number per thousand of the population per year. The difference between the two gives the *natural increase*, but the actual increase depends on migration. If there is a net migration into or out of a region, that will affect the actual rate of population change.

The *structure* of a country's population is shown by a graph called a population pyramid. It indicates the numbers of males and females in different age groups. When a population is increasing rapidly, the pyramid graph has a broad base. This indicates that there is a very large number of young people and relatively few old people. This kind of structure comes about when natural increase is high. In a country where the population is stable, the shape is quite different and shows roughly equal numbers in the age groups below sixty. This shape is common where birth rates and death rates are low and life expectancy is high.

Both types of population structure create problems. The broad base indicates a young population which has to be fed, clothed, housed and educated and in coming years will need jobs. The other pyramid shows a large and increasing number of elderly and mainly retired people which has to be supported by a small working population.

Migration of people affects total numbers. In the past, *international* migrations were very important. In the nineteenth and early twentieth centuries, millions of people migrated from Europe to North America. Today, *internal* migration is much more important, with movement of people from one part of a country to another. In the USA, for example, there has been a considerable movement of people from the north and east of the country to the south and west. On the countrywide scale the most important movement has been, and still is, from the countryside to the cities. This took place in Europe during the last few centuries, but more isolated regions continued to suffer from *rural depopulation* until quite recently, such as the Highlands of Scotland and the Massif Central of France.

In Africa, Latin America and Asia, migration from the countryside to the cities is taking place at a rapid rate. This *rural to urban* migration, combined with high natural growth rates, means that some cities are doubling their populations in ten years. As well as creating enormous cities, this is changing the balance of population between rural and urban areas. Just as the populations of Europe and North America are largely urban, so the rest of the world is moving in the same way. The level of *urbanisation*, as it is called, states the percentage of the population living in urban areas. Some idea of the changes that have taken place can be seen from the table (*right*).

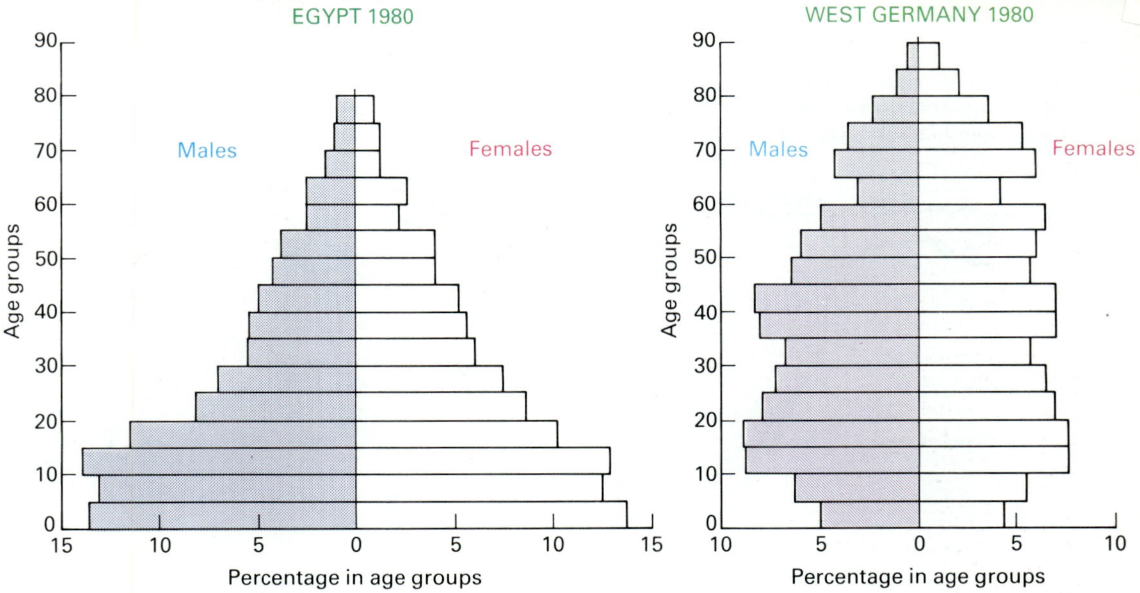

Figure 23.3 Population pyramids

The number of very large cities in the world has grown as a result of urbanisation. Not only are there many 'million' cities but the 'multi-million' cities with over $2\frac{1}{2}$ million people have increased in number rapidly. Until relatively recently, almost all the really large cities were in temperate latitudes in Europe and North America. Now, however, the Tropics and Asia, Latin America and Africa have most.

Country	*Percentage in urban areas* *1960*	*1980*
Algeria	30	52
Zaire	14	30
Canada	70	76
Mexico	50	66
Peru	47	67
India	15	22
Brazil	45	64
Netherlands	80	88

Figure 23.4 Developed and less developed countries

The map has been drawn using the Peters map projection. It shows areas correctly, being an equal-area projection. However, shapes are considerably distorted. Trace a set of world maps from atlases which use different map projections, such as Mollweide and Mercator. Compare the shapes and sizes of various parts of the world.

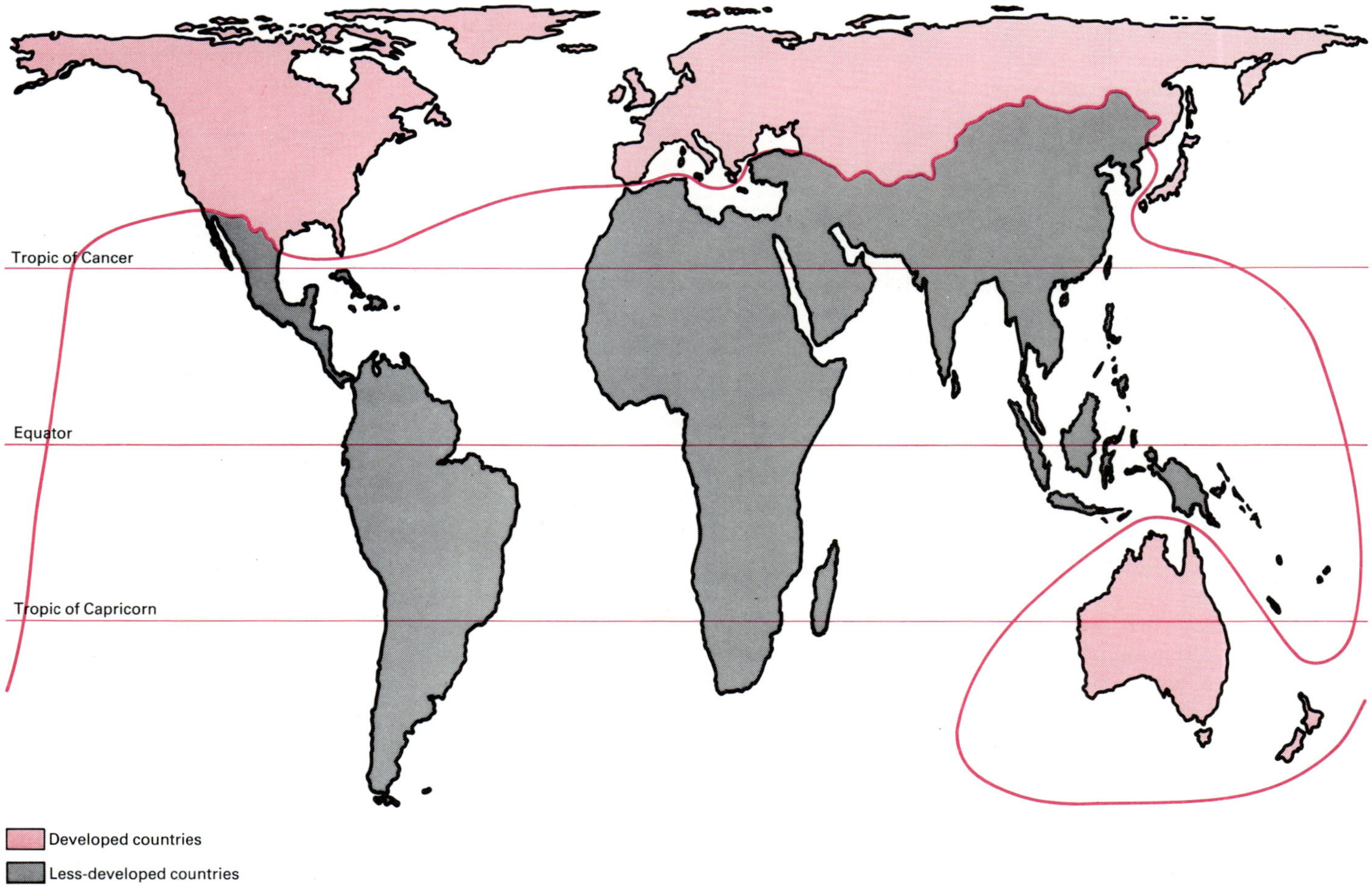

DEVELOPMENT

The distribution, growth and movement of population only deal with certain aspects of the geography of people. Another very basic feature is *wealth*. This has a great bearing on the quality of life. In general, those with greater wealth live longer, have a better education, live in better housing and have better medical services. A common measure of wealth is *gross domestic product* (GDP) per person. This is the total value of all the goods and services produced in a country divided by the number of people. This gives a means of comparing countries and can be used to identify rich, poor and very poor countries. The figures do not say how wealth is distributed within a country. It could be evenly distributed, or most of the wealth could be in the hands of a few people.

A variety of terms have been used to describe the richer and poorer countries of the world. The richer countries are usually called *developed countries* (DCs), the poorer *less developed* (LDCs) and the poorest *least-developed* (LLDCs).

Some indicators of development

	More developed	*Less developed*
GDP	high	low
Death rate	low	high
Birth rate	low	high
People per doctor	low	high
Literacy	high	low
Energy consumption	high	low
Employment in agriculture	low	high

Wealth alone does not give a full picture, and a wide variety of other sets of information are used as *indicators of development*. When they are used it is possible to see a broad relationship between the different features.

If actual statistics are used, it is possible to show the relationship between any two indicators on a scattergraph.

Sets of statistics show the situation at one time. Over time changes occur and the most important change has been the widening of the gap between the poorer and richer countries. The rich countries have become richer at a much faster rate than the others and the very poorest have hardly increased their wealth at all. There are, however, two groups of developing countries which have become richer. They are the oil-exporting countries like Kuwait and Saudi Arabia, and those in which industry has developed rapidly, like South Korea and Singapore.

Exercises

1 Write down reasons to explain why people migrate. Put your reasons into two groups: one for the reasons why people would want to leave a place, and the other for the reasons that would attract them to another place.

2 On an outline map of the world shade the regions which have high growth rates (2% and over). How does your map compare with Figure 23.5?

3 (*a*) On a map of the world mark and name multi-million cities (cities with over $2\frac{1}{2}$ million people). For the information, study a reference book.

(*b*) Comment on the distribution pattern shown.

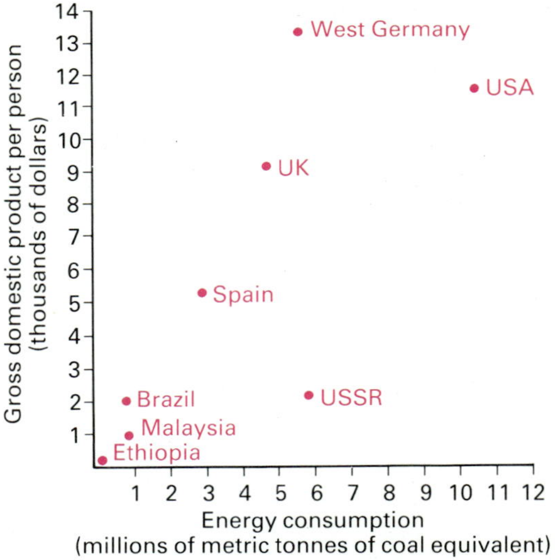

Figure 23.5 Scattergraph to show the relationship between wealth and energy consumption

World Population 1980

	Population (millions)	*% growth p.a.*
World	4432	1.7
North America	248	1.0
Latin America	364	2.5
Africa	470	2.9
East Asia	1175	1.4
South Asia	1404	2.2
Europe	484	0.4
Oceania	22.8	1.5
USSR	265	0.9

Chapter 24

Industry and resources

There are many types of industry. Some provide us with food and raw materials, for example, agriculture, forestry, mining, quarrying and fishing; these involve the direct exploitation of natural resources and are called *primary* activities. Others are concerned with the conversion of raw materials into useful commodities and are termed manufacturing industries or *secondary* activities. The transport and marketing of products, banking and insurance systems and all the other essential elements of a modern country like education and health systems are called services or *tertiary* activities.

ENERGY

An essential requirement for most industries is an adequate supply of energy. The most important primary sources of energy are coal, oil and natural gas, water and uranium. Alternative sources using the Sun's rays, wind and the tides are being developed. Electricity is not a *primary* source of energy. It has to be generated, and in the United Kingdom most is produced by steam turbines using coal or oil. Gas and uranium can also be used. Two-thirds of the world's electricity is derived from thermal sources, most of which originates from coal. Water-driven turbines are important in particular areas.

Coal

Coal is an organic rock which has been formed from vegetable matter. The black coals which occur in the United Kingdom were formed over 200 million years ago in the Carboniferous period. At that time vast stretches of the country were low-lying and marshy, with a dense vegetation which included tall plants similar to the present tropical tree ferns, giant horsetails and many smaller types. The dead matter accumulated until sea level rose, and sand and mud covered the organic layer. As the sea level fluctuated, numerous seams of coal were laid down with other rocks between them. The whole formation is known as the *Coal Measures*.

Since then other rocks have covered the Coal Measures, and earth movements have caused folding and faulting, distorting the coal seams. As overlying rock has been eroded, the coal has been exposed. Seams can vary in thickness from a few centimetres to 10 metres or more. Most coal is mined from hidden seams, but some is mined by open-cast methods, by removing overlying soil and rock and exposing the coal seams.

Coals formed more recently than the Carboniferous include *peat*. This is decayed vegetable matter, only partially decomposed. It has a small heating capacity but has been used in power stations. *Lignite* and *brown coal* are an intermediate type. They have a greater heating capacity than peat but less than black coal. Both are used in power stations and industrial concerns, for example, in Germany and the USSR.

Even Carboniferous coals can vary considerably. *Anthracite*, which has been subjected to great pressure, is a very hard coal, burns slowly with little flame or smoke, and gives great heat. Other black coals are generally grouped together as *bituminous coals*, but there are many grades of these, including steam coal, coking coal and household coal.

Most industrial areas in the United Kingdom

Figure 24.1 Chief sources of energy

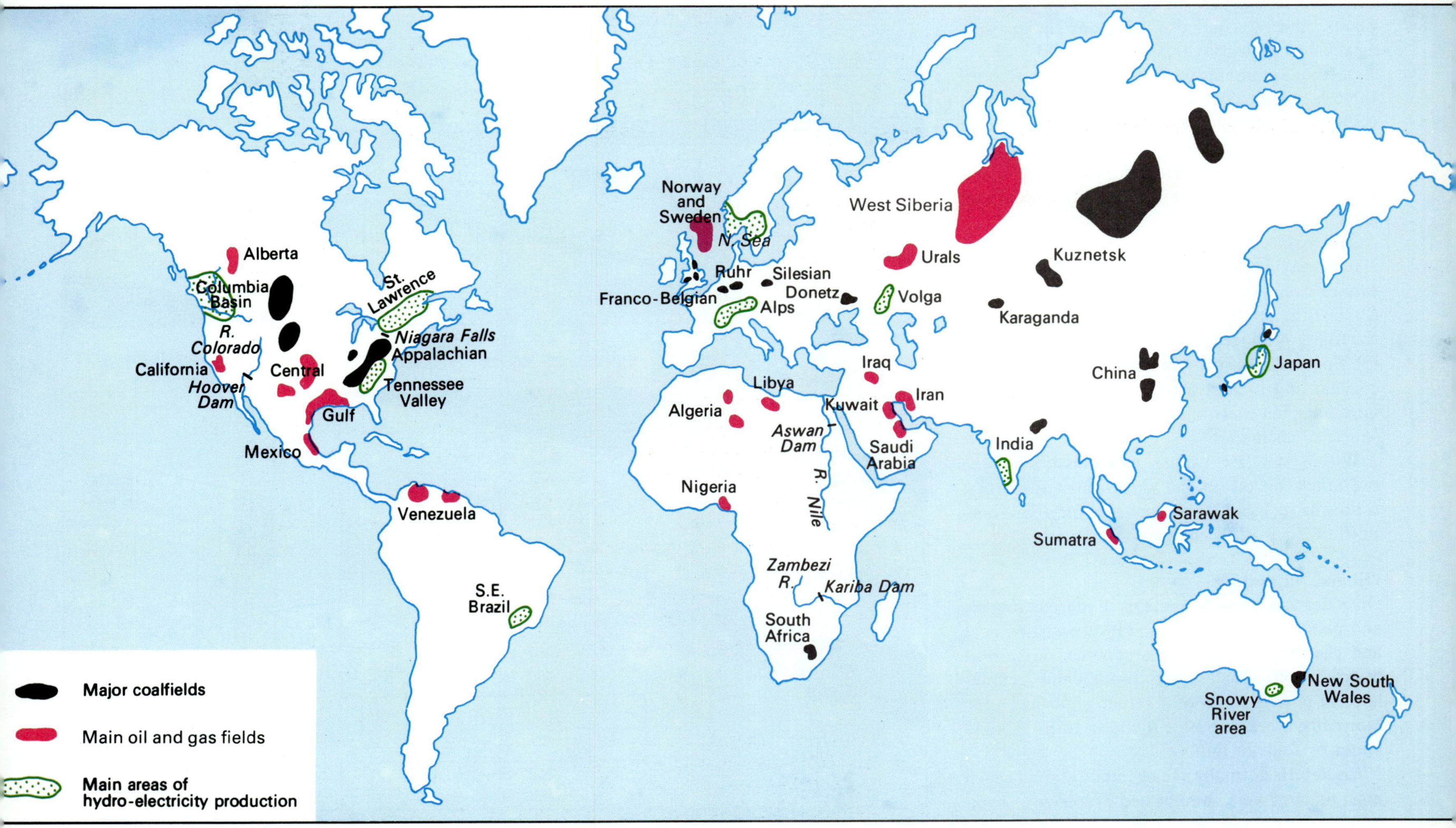

were developed on coalfields, relying on steam to drive machinery and on coke for iron and steel plants; this is true also of many other well-established industrial areas of the world.

Methods of mining vary considerably, but today there is a high degree of mechanisation in most coalmines, and the coalfields where expansion of mining is taking place are those most suited to mechanised production. For this, thick seams unbroken by faults and with only a little folding are preferred.

The main uses of coal are:

(*1*) In heavy industry, especially in the manufacture of iron and steel;

(*2*) For the production of thermal electricity, i.e., electricity produced from steam-driven turbines;

(*3*) For a variety of by-products including coke, coal tar and its derivatives, nylon, dyes and disinfectants, but coal is now second to petroleum as a source of chemicals.

Within the last twenty years coal has been overtaken by oil as the single most important source of fuel, but it still accounts for about one-third of the world's energy consumption.

Oil and Natural Gas

Oil is derived from the remains of marine animals and plants, probably as a result of pressure, heat and chemical action. It is a complex mixture of hydrocarbons which seeped through and accumulated in porous strata of different geological ages. Normally, workable oil is found in traps brought about by folds or faults.

Crude oil is normally a thick, muddy liquid which must undergo very complicated processes of *refin-*

Figure 24.2 Section through a coalmine

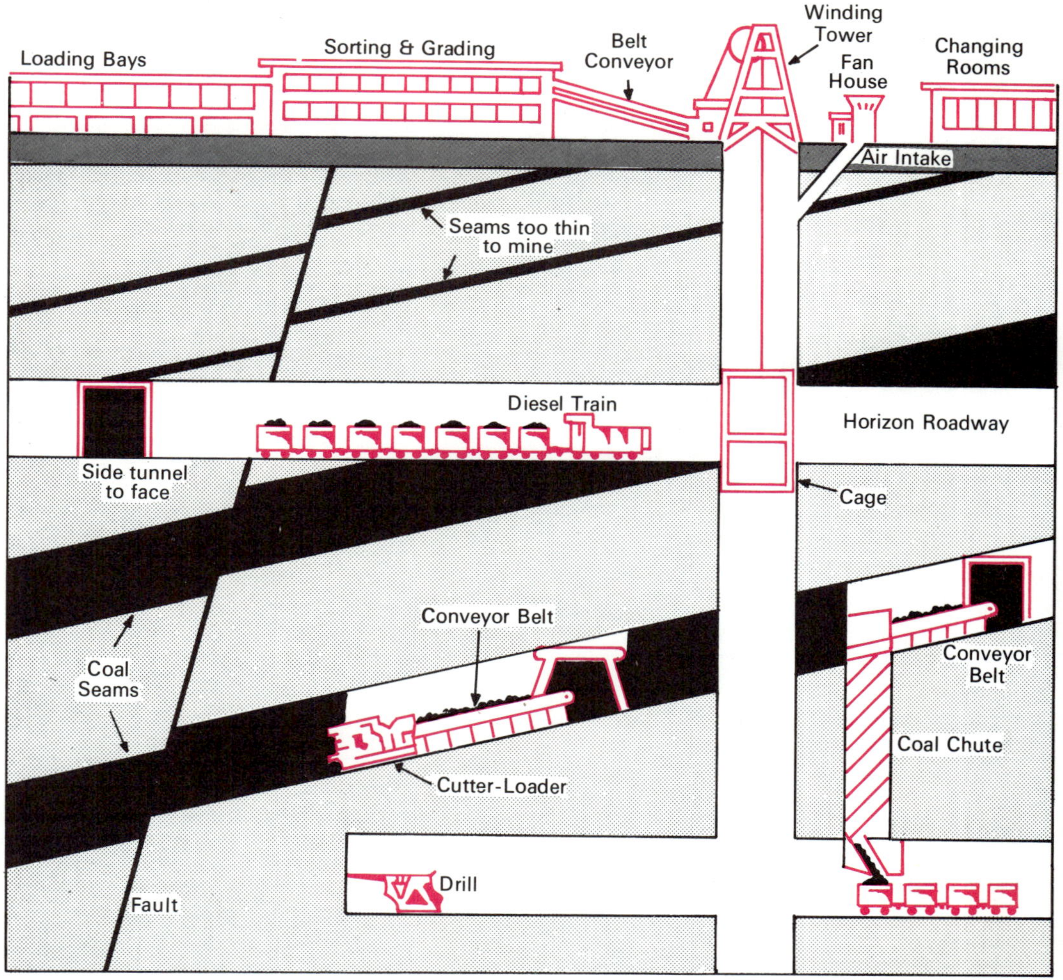

ing to separate it into its different grades. This is done by heating the liquid and then condensing the various vapours that come from it. The products from oil are many, and they can be grouped most conveniently as follows:

(*1*) Fuels, including petrol, paraffin and diesel oil;
(*2*) Lubricants and greases;
(*3*) Solids, which are often residual or base material after the refining has taken place; these include asphalt, paraffin wax, petroleum jelly and carbon black;
(*4*) By-products including anaesthetics, explosives, insecticides, synthetic rubber, fabrics and plastics.

Undoubtedly the most significant part oil has played has been in the development of transport. As petrol, diesel oil, heavy fuel oil and aviation fuels it powers almost all land, sea and air transport.

A big advantage that oil has over coal is that it is much easier to transport. It can be pumped from well to tanker and from tanker to refinery, or be carried vast distances by pipeline. Apart from the USA and the USSR, most major producers are in less-developed regions and export the bulk of their production. The oil makes such a great difference to their economies that oil-exporting countries are usually treated as a separate group.

The major producing areas are:

(*1*) The USA, the largest single producer, which exports little and needs to import some crude oil as it consumes 30 % of the world's oil;
(*2*) The USSR which has vast reserves, much of which are only now being tapped;
(*3*) The Middle East and North Africa which produce nearly 40 % of the world's supply;

Figure 24.3 A block diagram of an oilfield

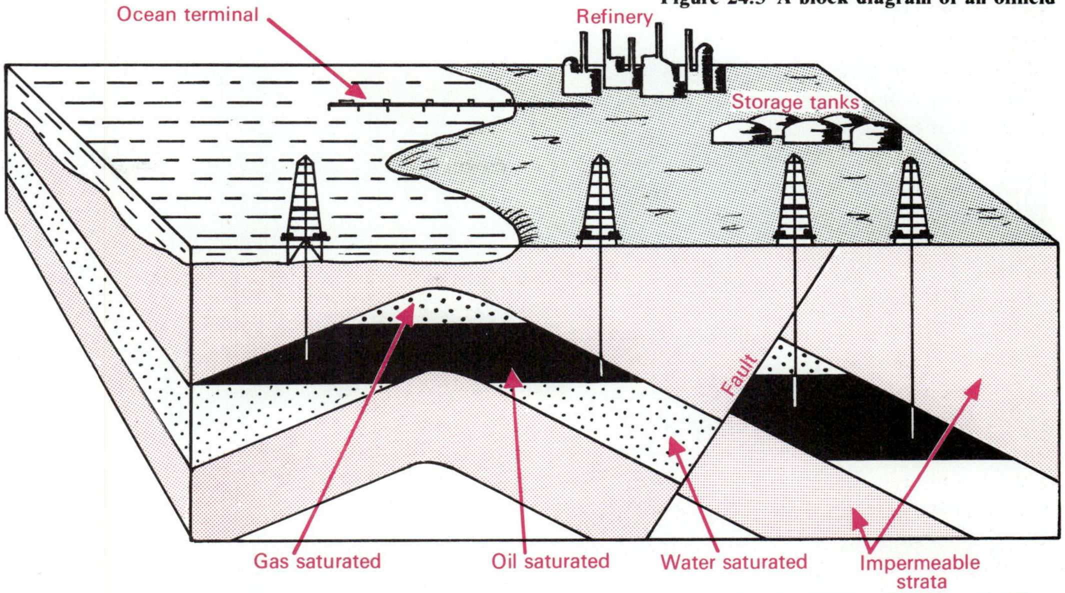

(*4*) Latin America, especially Venezuela and Mexico, is the other major producer;
(*5*) Outside these regions are individual countries such as Nigeria, the United Kingdom, Indonesia and China, which are oil producers on a major scale.

The location of *oil refineries* has been linked to the major markets. Therefore most have been located in the industrial countries. Locations have been more restricted by the need for deep-water harbours to accommodate very large oil tankers. Petrochemical industries, using by-products of crude oil, have developed near most refinery locations.

Natural gas is often associated with oilfields. It is highly valued as a clean and convenient fuel and, like oil, can be piped over considerable distances. For example, natural gas from the West Siberian fields in the USSR is piped to Western Europe. It is used in homes and in factories, and has almost entirely replaced gas derived from coal.

Hydro-electric Power

Hydro-electricity provides about 2 % of the world's energy needs. Hydro-electric power stations use the force of falling water to drive turbines which in turn drive generators. The ideal site is one where a large volume of water falls through a considerable height, with little variation in flow from season to season. Such conditions rarely occur in nature, but construction work can create them.

Natural falls can be developed and these tend to be in mountain areas. As a result the bulk of HEP schemes are located in mountainous areas in North America and Europe. The enormous amount of money needed for such schemes means that the HEP potential of developing countries has not, on the whole, been utilised. Notable exceptions are S. E. Brazil and individual schemes such as the Aswan High Dam in Egypt and the Kariba Dam on the Zimbabwe/Zambia border.

There are two distinct types of man-made falls. The first is the *diversion system* where water is led off from a river by a canal so that the fall of water at a particular point can be increased. The second is the *dam*. This needs a suitable catchment area to permit the collection of water behind the dam. A steep-sided valley provides the best setting. To enable power stations to be built in more accessible locations, water is sometimes led through tunnels for several kilometres from lakes. This reduces the problems of building electricity transmission lines across mountainous terrain.

Renewable and Non-renewable Resources

Hydro-electric power does not use up resources which cannot be replaced. As such it is different from coal, oil and gas. It illustrates the difference between *renewable* resources, which include, for example, HEP, forest reserves and solar energy, and

Beauchastel on the River Rhône, France (*Photo E D F*)

Two generating stations on the Niagara River (*Ontario Hydro*). Compare this scene with that of the system at Beauchastel. Use diagrams to help you explain the differences.

non-renewable resources such as fossil fuels and mineral deposits, which cannot be replaced. There is an overlap between the two sets of resources, because the way a resource is utilised affects its long-term future. Overfishing, for example, has led to the collapse of the herring industry, and the total extinction of some species of whales means that no replenishment of that resource is possible.

With regard to energy resources, there is a greater awareness now of the non-renewable nature of the main sources. Although coal reserves are sufficient for hundreds of years, the situation with oil is far less happy. It is for this sort of reason that greater efforts are being made to *conserve* energy or to make more efficient use of it.

Nuclear Energy

Nuclear energy is obtained primarily as heat from a reaction involving the *fission* or splitting of atoms of certain elements, of which uranium is the most important. Only small amounts of fuel are needed, and uranium is widely distributed, with the USA, South Africa, Canada and the USSR being the major producers. Just over 2% of the world's energy consumption is from this source, and that is mainly in the developed countries. A major problem is the disposal of waste and public fears about the safety of some types of nuclear reactor.

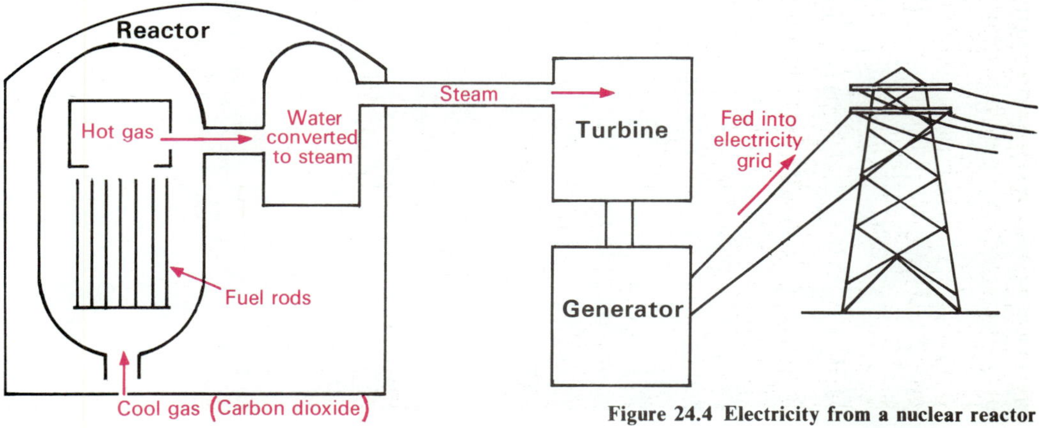

Figure 24.4 Electricity from a nuclear reactor

Hamilton, Ontario is a large steel producing centre in Canada

Study the photograph. What factors do you think led to the establishment of a steel works here?

RAW MATERIALS

Many animal and vegetable raw materials have already been considered in the various chapters on the natural regions of the world. Many other raw materials are of mineral origin and will now be considered.

All minerals fall into the category of non-renewable resources. Worldwide, there are large reserves of most minerals, but whether or not deposits are counted as reserves depends on their quality, their price and the state of mining technology. In addition, many areas of the world have yet to be geologically surveyed in detail.

There are several points to be made about the location of mining:

(*1*) Large mineral deposits are usually mined before small ones;

(*2*) Rich deposits with a high mineral content are mined before poorer ones;

(*3*) Accessible deposits are mined before less accessible;

(*4*) Easy-to-mine deposits are developed before those presenting mining difficulties;

(*5*) Rises in prices or falls in prices bring into production or take out of production marginal mines;

(*6*) Exploration for and development of one mineral often uncovers the presence of other minerals. In combination they may be worth mining.

Most metallic minerals of economic value are found in mountainous areas or shield areas. Among the first group are the Western Cordillera of North America, the Andes, the Urals and the fold mountains of Siberia. Among the second group are the Canadian Shield, Scandinavian Shield, Brazilian Plateau and the Zaire–Zambia region. Most of these areas are remote from large industrial centres where minerals are required.

In many cases the metallic minerals were deposited when molten lava intruded into the upper strata of the Earth's crust. The ores in this lava solidified in cracks (veins) or hollows (lodes or cavity fillings) within the sedimentary rocks, or impregnated them (replacement ores). Sometimes the rocks in which the ore occurred were eroded, so that the ore was removed. This led to the widespread alluvial deposits of many metal ores.

When ore is mined there is always a large proportion of waste. Some iron ores are mined when there is under one-fifth metal content; with other metals the proportion of waste is often much greater, and with some rare metals it may be over 99%. It is often necessary to refine or partly refine the ore before sending it to the markets, otherwise one would be paying more for the transport of waste than for the useful metal.

Minerals are usually considered in different groups.

Iron Ore and Ferro-Alloys

Iron ore is, of course, the raw material from which iron and steel are made, and it is the most widely used metal ore. There are various types of iron ore, including magnetite (with up to 68% metal content), haematite (up to 65% metal) and poorer grade ores such as limonite and siderite. In the past the steel industry developed in locations where both iron ore and coking coal were found adjacent to each other. Now, however, the major deposits being worked are often in isolated areas of the world and the iron ore is transported great distances. For example, most of the iron ore mined in Western Australia is transported by sea to Japan, and iron ore from Brazil and West Africa is transported to Europe. Major reserves of iron ore are found in North America, especially in northern Canada, in the USSR, Brazil and West Africa.

Ferro-alloys are the metals which are added in the manufacturing process to give steels of different types. They include nickel and chromium which give resistance to corrosion, tungsten and vanadium to give hardness, and molybdenum to give elasticity.

Non-Ferrous Metals

In this group are five very important metals, each of which has valuable industrial applications. The bulk of the world's *tin* comes from South-East Asia, where it is generally mined from alluvial deposits, and Bolivia, where there are deep mines high in the Andes. *Copper* is produced in five main areas, the Western Cordillera of the USA, the Copper Belt of Zambia and Zaire, the Andes in Chile and the Urals and Siberian areas of the USSR. *Lead and zinc* are usually found together. The Western Cordillera of North America, from Canada to Mexico, is one main production area, with the USSR and Australia also being major producers. *Aluminium* is generally obtained from the mineral bauxite, and it is found in completely different locations from the other minerals. The leading producers are Australia, Guinea, Jamaica, USSR, Brazil, Yugoslavia and Surinam. Since the manufacture of aluminium requires a large amount

Figure 24.5 Major mineral deposits and steel-producing areas

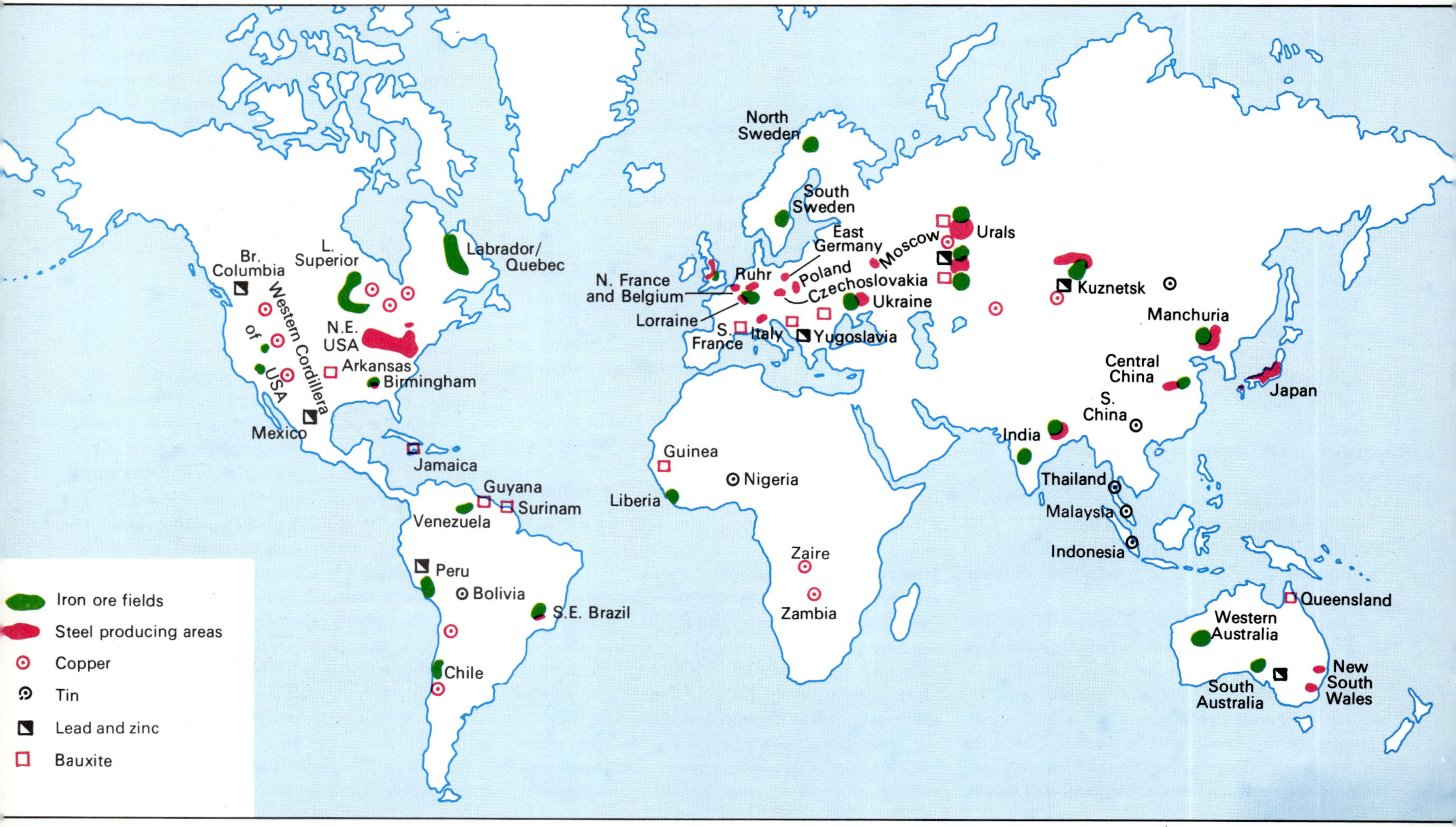

of cheap electricity, it is frequently produced in areas a long way from the bauxite source. This is often in areas of great HEP development, such as British Columbia and Quebec.

Precious Metals

The main sources of *gold* are deep mines in South Africa, Canada, the USA and the USSR. *Silver* is obtained from lead, zinc, copper and gold ores. *Platinum*, a very rare mineral, is obtained from Canada, South Africa and the USSR.

Non-Metallic Minerals

These are more generally available and include *clays* of various types and qualities for use in the pottery industry and in the production of bricks and tiles. *Chalk* and *limestone* are the main raw materials in the production of cement, and *sand* and *gravel* are also important in the construction industry.

Minerals for the Chemical Industry

The chemical industry covers the manufacture of a very wide range of products including plastics, synthetic fibres, dyes, drugs, disinfectants, soaps and detergents, cosmetics and paints. In addition to the various by-products of coal and oil, there are three other important raw materials. These are *sodium chloride* (common salt), *sulphur* and natural *nitrates*.

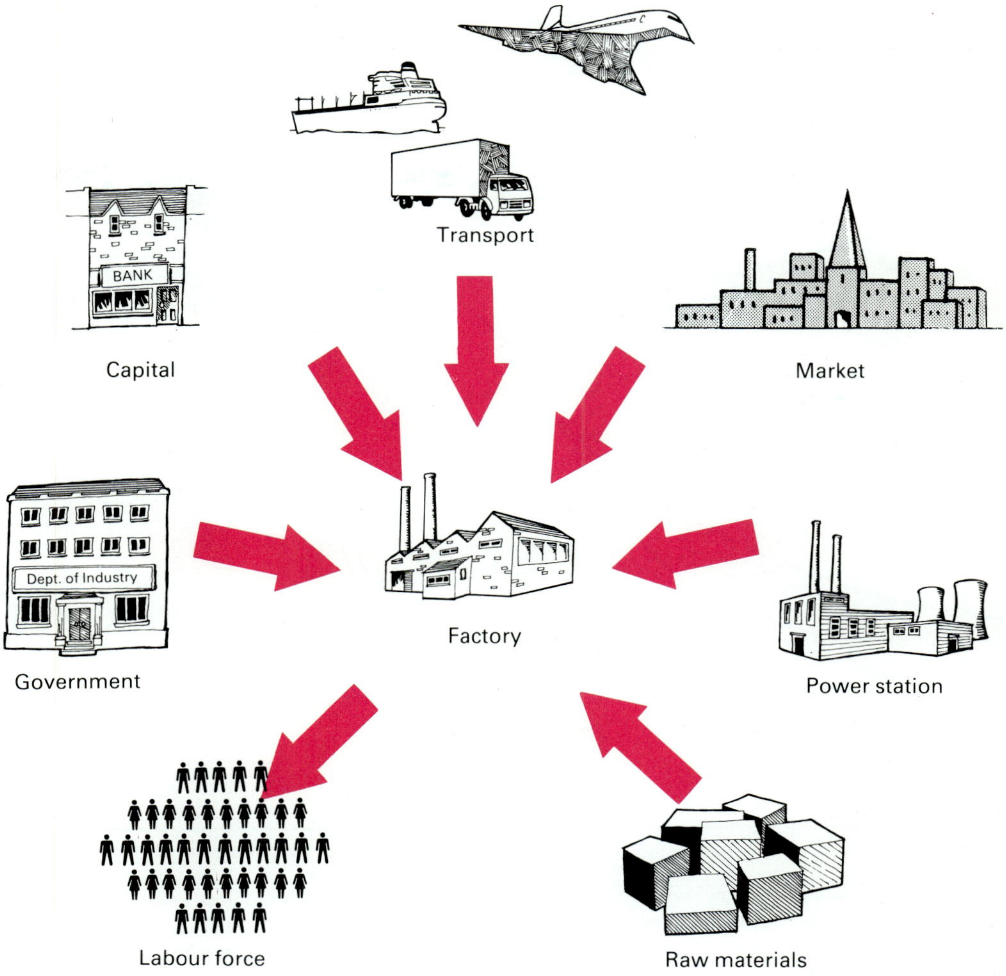

Figure 24.6 Factors influencing the location of industry

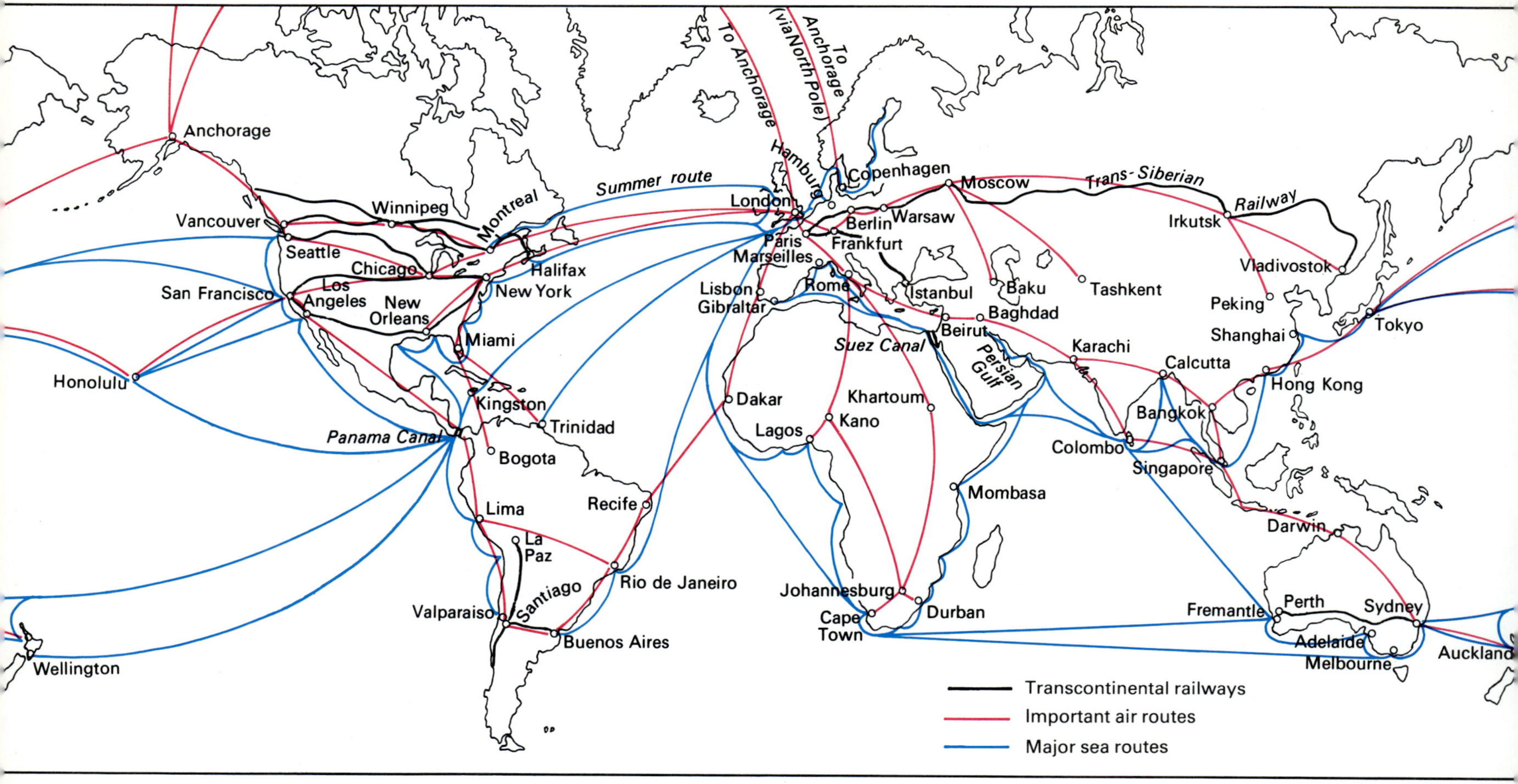

Figure 24.7 Major world routes

(*a*) Describe three sea routes between the United Kingdom and Sydney (Australia), naming two ports of call on each route and any two commodities which a ship might pick up at each port.

(*b*) Describe three air routes between London and Tokyo (Japan). Which is the shortest route? What types of cargo are generally carried by air?

(*c*) Name a sea route on which on which you would expect to find each of the following: oil tankers, iron ore ships, timber ships, grain ships and banana ships.

(*d*) Name three countries (in different continents) which were formerly ruled by colonial powers. State their previous names and the name of the colonial powers.

(*e*) How does the map help to explain why Britain is a great trading nation?

MANUFACTURING INDUSTRY

Manufacturing industry varies from basic raw material processing to the production of highly sophisticated technological equipment. It includes the manufacture of transport equipment, such as shipbuilding and the motor industry, textiles, chemicals and many new lighter industries, including computers and electrical engineering. Many of the newest types of industry depend on the results of scientific research.

The pattern of industry is not fixed, and recent years have seen the rapid growth of industry, not only in Japan but in other countries not previously noted for their industry. Spain is an example of such a country in Europe. In the developing world, industry has grown in what are called the 'newly-industrialising countries', a group that includes South Korea and Singapore, for example. In developed countries there has been a decline in many basic industries such as iron and steel, and newer 'high-technology' industry tend to locate in very different areas from the old basic industries. In the USA, for instance, the South and South-West of the country is where these growth industries tend to locate. In the developed world, too, a process called de-industrialisation has been forecast, with greater emphasis on service and research-based activities, with actual manufacturing becoming more and more automated and employing fewer people.

Exercises

1 On an outline map of the world:

(*a*) Mark in the main coalfields as shown in figure 24.1.

(*b*) Name the countries which have important fields.

Coal production 1982 (thousand tonnes)

USA	756 060
China	600 000
USSR	510 000
Poland	189 360
South Africa	136 908
India	128 304
United Kingdom	124 656
Australia	96 792
West Germany	89 792
Czechoslovakia	27 240
Canada	22 296
Total world	2 861 000

(*c*) Against each country insert the percentage of world production (to the nearest whole number), calculating this from the statistics.

2 On an outline map of the British Isles:

(*a*) Mark the coalfields.

(*b*) Mark and name oil refineries, hydro-electric schemes and nuclear power stations. Refer to an atlas.

(*c*) Mark with a dot and name all the towns with a population of over 200 000. You may find the information you require in an atlas or a reference book such as the *Statesman's Year Book* or the *Geographical Digest*.

3 Study the following statistics:

Iron ore production 1982		*Crude steel production 1982*	
(thousand tonnes)			
USSR	146 397	USSR	147 000
Australia	55 188	Japan	99 552
Brazil	42 652	USA	66 132
China	32 500*	West Germany	36 348
India	25 688	China	27 000*
USA	23 731	Italy	23 940
Canada	20 151	France	18 408
South Africa	15 959	Czechoslovakia	15 036
Liberia	13 398*	Poland	14 474
Sweden	10 006	United Kingdom	13 704

*estimated

Plot these statistics graphically, and after studying the graphs state which countries are likely to be (i) leading importers of iron ore, (ii) leading exporters.

4 Make a list of products used in your home, stating their use, the name of the raw material used if appropriate, and the country in which the product was made.

5 Make a study of a factory near to your home and write an account to include the following: (*a*) the products, (*b*) the source of energy, (*c*) the raw materials or components used and from where they are obtained, (*d*) the numbers and type of workers, i.e. skilled or unskilled, male or female, (*e*) the processes of manufacture, and (*f*) the principal markets and methods of distribution.

Index

Abrasion by ice, 89
Abrasion by wind, 82
Acacia, 118
Agribusiness, 159
Alluvial plain, 71
Aloe, 126
Alpacas, 178
Alps, 178
Aluminium, 195
Amazon basin, 111
Andes, 178
Anemometer, 49
Anticyclones, 53
Apples, 145, 159, 161
Apricots, 144, 145, 160
Arabian desert, 127
Arabs, 127
Arêtes, 89
Artesian well, 83
Asbestos, 197
Aswan dams, 129
Atacama, 129
Atmosphere, 40–46
Atoll, 108
Aurora, 38
Australian desert, 129

Badland, 83
Bamboo, 134
Bananas, 119, 122
Baobab, 118
Barchan, 83
Barley, 152, 153, 159
Barometer, aneroid, 48
  mercury, 48
Bauxite, 195
Beach formation, 96
Beaufort scale, 50
Bedouin nomads, 127
Beef, 113, 148, 150
Bessemer process, 193
Bison, 153
Black earth region, 152
Boulder clay, 94
Bridges, 11
Brown coals, 188
Butte, 84

Cactus, 126
California, 145
Camels, 127, 153
Campos, 121
Canning, 145
Cassava, 119
Cattle—dairy, 159, 166
Cattle rearing, 119, 121, 122, 149, 150, 152, 153, 159
Chalk scenery, 81
Cheddar Gorge, 80
Chemical industry, 197
Chernozems, 152
Cherries, 159
Chocolate, 114
Cirques, 89
Citrus fruit, 144
Clay, 197
Cliff coasts, 97
Climate graphs, 51
Climate maps, 51
Climatological stations, 47, 48
Cloud, 41, 42
Coal, 166, 188
  formaton, 188
  mining, 190
  resources, 189
  types of, 188
  uses, 190
Coalfields, 165, 184
Coastal plain, 101
Coastlines, longitudinal type, 99
  transverse type, 97

Cocoa, 113
Coconut palm, 116
Cod, 160, 166
Coffee, 122
Cold temperate regions, 170
Collectives, 152
Coir, 116
Communications, railways, 11
road, 11
water, 10
Compass points, 3
traverse, 4
Condensation, 41
Coniferous forest belt of Canada, 171
Coniferous forest belt of Eurasia, 174
Coniferous trees, 157, 161, 164, 170, 171, 172, 174
Continental shelf, 103
Contours, 4
Convectional rainfall, 42, 110
Cool temperate eastern margins, 164
of Canada and the USA, 164
of Manchuria and E. Siberia, 166
Cool temperate western margins, 157
of Australasia, 161
of North America, 160
of North West Europe, 158
of southern Chile, 161
Coral, 107
Corona, 35
Cotton, 119, 122, 127, 129, 147, 148, 153
Copper, 195
Copra, 116
Cork oak, 140
Crag and tail formation, 95
Cross sections, 6
Cyclones, 42
Cyclonic rain, 42, 157

Dairy products, 145
Dalmatian coast, 102
Date palm, 127, 130
Deccan, 139
Deciduous trees, 118, 133, 146, 157, 164
Deforestation, 114, 133
Delta, 75
Depressions, 42, 52, 53, 157
Desertification, 86
Deserts, types of, 82
Dew, 44
Dew point, 41
Diamonds, 129, 197
Direction on maps, 2, 3
Doldrums, 40, 110, 118
Donkeys, 127
Douglas fir, 161, 172
Drumlins, 94
Dry farming, 153
Dust bowl, 86

Earth, its movements in space, 30–35
Earth movements, 59–69
Earthquakes, 60
Earth's crust, 58, 59
East Indies, 115
Eclipse, lunar and solar, 35
Energy, 188–193
Equatorial Africa, 113
Equatorial regions, 110
Equinoxes, 32
Erosion, lateral, 71
vertical, 70
Erratics, 94
Eskers, 94
Eskimos, 179
Estancias, 150
Estuaries, 59, 74, 102
Euphorbia, 126
Euphrates River, 129
Evaporation, 41
Evergreens, 110, 140, 146, 170
Extensive farming methods, 10, 153

Farming—intensive, 10
Faulting, 63
Fazendas, 122
Fellahin, 127
Figs, 144
Fiords, 75
Fishing, 160, 166, 167, 179
Fishing grounds, 106
Flood plain, 71
Fog, 43
Fold mountains, 62
Fur bearing animals, 171
Fur farms, 171, 174

Ganges basin, 134
Garrigue, 140
Geological history, 59
Geysers, 67
Glaciation in mountain areas, 90–93
in lowland areas, 94, 95
Goats, 113, 127, 144, 145, 154
Gold, 129, 197
Gorges, 74
Gradient, 6
Grand Banks, 166
Grand Canyon, 85
Grapefruit, 144
Grapes, 144, 145
Great circles, 33
Greenwich Mean Time, 33
Groundnuts, 119, 139
Guinea corn, 119

Hachures, 4
Halibut, 160
Hanging valleys, 90
Hardwoods, 110, 114, 133, 157

Herring, 160
Hides, 119
High mountain areas, 175
Hill shading, 4
Hoar frost, 44
Hoover dam, 129
Horse latitudes, 140
Horst, 63
Hot deserts, 126
Humidity, 41
Humidity, measurement of, 48
Hurricanes, 42
Hydro-electric power, 166, 171, 174, 191, 192
Hydro-electricity, 174
Hydrological cycle, 45
  in a river basin, 76
HYV, 136

Ice Age, 90
Icebergs, 89
Ice sheets, 89, 94
Igneous activity, 63–69
Igneous intrusions, 65
Incised meanders, 76
Indonesia, 115
Indus Plain, 139
Indus River, 129, 139
Ingleborough, 79, 80
International date line, 33
Ionosphere, 38
Iron and steel industry, 195
Iron ore, 195
Iron ore fields, 166, 191
Irrigation, 127, 136, 139, 141, 144, 145, 153, 161
  basin, 129
  perennial, 129
  tank method, 139
Isobars, 51
Isohyets, 51
Isotherms, 51

Jute, 136

Kalahari, 129
Kiaolung, 166
Kirghiz, 153
Krakatoa, 63

Labrador current, 104, 164
Laccolith, 65
Lamb, 161
Land and sea breezes, 41
Land forms, definitions, 4, 5
Land use, 10
Lapps, 180
Latex, 115
Latitude, 6, 7, 33
Layer colouring, 4
Lead, 195
Lemons, 144, 145
Levees, 72
Lignite, 188
Limestones—for building, 197
Limestone scenery, 77–80
Linseed, 139
Llamas, 178
Llanos, 121
Loess, 82
Log book, weather, 50
Longitude, 6, 7, 33
Longshore drift, 99
Lumbering, 166, 171–174

Magnetic compass, 3
Magnetic variation, 3
Maize, 119, 129, 134, 139, 144, 148, 149, 152, 154, 166
Malaysia, 115
Malham, 80
Manchuria, 166

Mangrove swamps, 110
Manioc, 113
Map references, 7
Maquis, 140
Market gardening, 159, 166
Market towns, 9
Meanders, 71
Mediterranean area of California, Central Chile, New Zealand, South Africa, Southern Australia, 145
  regions, 140
  Sea area, 141
Mesa, 84
Mesopotamia, 129
Metals non-ferrous, 195
  precious, 197
Meteorological stations, 47–50
Millet, 119, 129, 134, 166
Mineral oil, 190–191
Minerals—world distribution, 192
Mist, 43
Monoculture, 153
Monsoon forest, 133
Monsoon lands, 131
Moon, 30, 34, 35
Moraines, 89, 90, 94
Mutton, 154, 161, 167

National grid, 7
Natural gas, 190
New towns, 174
Niagara Falls, 73
Nile River, 128, 129
Nitrates, 129, 197
Nomads, 127, 153
North Atlantic Drift, 104, 158
Nuclear energy, 193

Oasis, 83, 127
Oats, 149, 152, 153
Ocean currents, 104

Ocean floor, 103
Oceans and seas, 103–108
Oilfields distribution, 189, 191
Oil palm, 114
Oil seeds, 139
Olives, 144, 145
Opisometer, 2
Oranges, 122, 144, 145
Orinoco River, 121
Orogenies, 62
Overflow channels, 95
Ox-bow lakes, 71
Oxen, 136

Palm oil, 114
Pampas, 150
Patagonia, 167
Peaches, 144, 145
Pears, 144, 145, 159
Peat, 188
Pigs, 113, 149, 154
Planets, 30
Plankton, 106
Plantains, 113
Plate boundaries, 68, 69
Platinum, 197
Plums, 145, 159
Polar front, 40
Potatoes, 145, 178
Pottery, 197
Prairies, 153
Pressure, atmospheric, 40
Pressure, atmospheric—measurement of, 40, 48
Public utilities, on maps, 10
Pulp and paper manufacture, 172

Rain gauge, 49
Rain, types of, 42
Raised beaches, 59
Rayon, 172
Reindeer, 179, 180
Relief on maps, 4
Rias, 74
Ribbon development, 9
Ribbon lakes, 90
Rice, 129, 134, 139
Ridge of high pressure, 53, 54
Rift valley, 63
River basin, 75
River terraces, 76
Rivers, 70–76
Roches moutonnées, 90
Rocks, types of, 58
Rotation of crops, 86, 159
Rubber, 113, 115

Sahara desert, 127
Salinity of oceans, 103
Salmon, 161
Salt, common (sodium chloride), 197
Salt lakes, 84
Salt marsh, 100
Samoyeds, 180
Sand dunes, 83, 101
Savanna regions, 118
  of Africa, 119
  of Australia, 121
  of South America, 121
Scale of a map, 2
Screes, 70
Scrub, 118, 127, 133
Sea coast, 96–102
Seals, 179
Seasons, 31, 32
Seif dunes, 83
Seismographs, 60
Selvas, 111
Sequoia, 161
Sesamum, 139
Settlement—Anglo-Saxon, 8
  factors favouring, 8, 9
  prehistoric, 8
  Roman, 8
Sheep, 127, 144, 145, 152, 154, 159, 161, 167
Shields, 62
Shifting tillage, 113
Silver, 197
Sisal, 119
Smog, 44
Snowdon, 90
Snow line, 89
Soft fruit, 159
Softwoods, 171, 172
Soil creep, 85
Soil erosion, 85, 153
Solar system, 30
Solstices, 32
Sources of power, 188
Soya, 147
Soya beans, 139, 147, 166
Spaghetti, 144
Spits and bars, 99
Spot heights, 4
State farms, 152
Steppes, 152
Stevenson screen, 48
Stratosphere, 38
Submerged forests, 60
Subsistence farming, 136
Suburbs, 9
Sudan, 119
Sugar beet, 152
Sugar cane, 119, 122, 129
Sukkur dam, 129
Sulphur, 197
Sunflowers, 152
Sunshine recorder, 50

Taiga, 174
Tea, 139

Teak, 133
Temperate interior regions, 151
  in the southern hemisphere, 154
Temperate monsoon areas, 139
Temperature, 39
  factors determining, 39
  inversion of, 44
  measurement of, 48
Temperature of oceans, 103
Tennessee Valley Authority, 86
Thermometer, grass minimum, 48
  wet and dry bulb, 48
Thermometers, maximum and
  minimum, 48
Tides, 106
Tigris River, 129
Time, 33, 34
Tin, 195
Tobacco, 119, 122, 148
Tourism, 11, 144, 179
Trade winds, 40, 118, 122, 126
Trans-Amazonica Highway, 113
Transhumance, 144, 178
Trapping, 171, 174
Triangulation station, 4
Tropical and sub-tropical east coasts, 122
Troposphere, 38
Truck farming, 166
Truncated spurs, 90
Tsetse fly, 119
Tundra—Eurasian, 180
  North America, 179
  southern hemisphere, 180
Typhoons, 42

Underground water, 76, 77

Vale of Pickering, 95
Veld, 154
Vertical interval, 4
Vesuvius, 65
Vicunas, 178
Visibility on maps, 6
Volcanoes, 63–65

Wadis, 83
Walnuts, 144
Warm temperate eastern margins, 146
  in S. Africa, 150
  in S. E. Australia, 150
  in S. E. USA, 147
Warm temperate western margins, 140
Waterfalls, 72
Water, in the air, 41, 42
Wave action, 96
Weather forecasts, 51
Weather graphs, 51
Weather maps, 51–54
Weathering, 70
Weathering in arid regions, 82
Westerly winds, 40, 140, 157, 164
Water table, 77
Whaling, 180
Wheat, 129, 139, 141, 145, 149, 150, 152, 153, 154, 159,
  166, 172
Winds, local, 41
Winds, prevailing, 40
Wind vane, 49
Wool, 154, 167

Yams, 113, 119

Zaire Basin, 113
Zebu cattle, 119
Zinc, 195
Weathering, 70
Weathering in arid regions, 82
Westerly winds, 40, 140, 157, 164
Water table, 77
Whaling, 180
Wheat, 129, 139, 141, 145, 149, 150, 152, 153, 154, 159,
  166, 172
Winds, local, 41
Winds, prevailing, 40
Wind vane, 49
Wool, 154, 167

Yams, 113, 119

Zebu cattle, 119
Zinc, 195